INDIGENOUS TECHNOLOGIES
FOR
DEVELOPED INDIA
NATIONAL SCIENCE DAY

DIPAN KUMAR DAS
SUDIP KUMAR DAS

On this auspicious National Science Day, we dedicate our collective efforts and unwavering commitment to the vision of a Developed India propelled by Indigenous Technologies. Today, we celebrate the spirit of scientific inquiry, innovation, and the invaluable contributions of our scientific community.

To the Visionaries who dared to dream, To the Innovators who turned dreams into reality, To the Researchers who delved into the unknown, To the Engineers who built the foundations of progress, To the Technologists who harnessed the power of creativity,

In the tapestry of India's journey towards development, your dedication to advancing Indigenous Technologies has been the guiding light. On this National Science Day, we honor your tireless pursuit of knowledge, your resilience in the face of challenges, and your passion for creating a better tomorrow.

May our Indigenous Technologies be the cornerstone of a Developed India — a nation where innovation knows no bounds, where sustainability is ingrained in our progress, and where the fruits of scientific endeavors benefit every citizen.

Happy National Science Day!

Foreword

In the ever-evolving landscape of progress, innovation emerges as the catalyst that propels nations into transformative realms. As we embark on a journey into the world of possibilities, it is with great enthusiasm that I introduce this compilation dedicated to the exploration of Indigenous Technologies and their pivotal role in shaping a Developed India.

In the pages that follow, we delve into the intricate tapestry of India's technological landscape, celebrating the convergence of tradition and modernity. Indigenous Technologies, rooted in the wisdom of generations, stand poised to lead the nation into an era of unprecedented growth, sustainability, and global influence.

Our narrative begins with an exploration of the challenges faced by innovators in securing funding, be it through government grants, private investments, or venture capital. We shed light on successful funding models, envisioning investment ecosystems that champion indigenous technological ventures. From economic growth to employment opportunities, the potential is vast, and the path to realizing it requires strategic collaboration and support.

Technological barriers often loom large on the innovation horizon. In addressing these challenges, we uncover solutions that bridge infrastructural gaps and expand access to cutting-edge tools. Our journey takes us into the realms of artificial intelligence and biotechnology, where Indigenous Technologies promise to revolutionize industries and contribute to the overall development of the nation.

Communication lies at the heart of progress, and our exploration extends to the challenges of communicating the value of technologies to the public. Strategies for bridging the awareness gap and the role of education and outreach programs take center stage, fostering a culture that embraces indigenous innovations.

As we shift focus to the opportunities that arise from embracing indigenous technologies, the narrative unfolds into the realms of economic growth, employment creation, and sustainable development. Our exploration extends to the potential for global collaborations and market expansion, envisioning a future where India's technological prowess influences the global stage.

Emphasizing the need to cultivate a culture of innovation and self-reliance, we examine how educational institutions, research centers, and industries can collaboratively nurture a mindset that values indigenous technologies. The chapter leaves readers with a vision of a future where innovation becomes ingrained in the nation's DNA, driving continuous progress.

Our gaze then turns toward the future, envisioning the trajectory of indigenous technologies in shaping the destiny of India. The promising landscape of emerging fields and the continued growth of traditional knowledge systems offer a glimpse into a future driven by innovation and technological prowess.

The integration of cutting-edge technologies becomes paramount in our narrative, exploring the potential of indigenous technologies in emerging fields, with a special focus on artificial intelligence and biotechnology. From AI-driven solutions to genetic advancements, the chapter delves into how these technologies can revolutionize industries and contribute to the overall development of the nation.

In acknowledging the wealth of wisdom passed down through generations, we highlight the symbiotic relationship between ancient knowledge and contemporary innovation. Traditional practices in agriculture, healthcare, and more serve as the foundation for harnessing the strengths of indigenous technologies.

Our journey extends to the significant contributions of indigenous technologies to sustainable development, addressing environmental challenges, and fostering eco-friendly practices. From renewable energy solutions to waste management innovations, the chapter envisions a future where indigenous technologies play a pivotal role in achieving balance and harmony with nature.

Nurturing the next generation of innovators takes center stage in our exploration. From education and mentorship to creating conducive environments for innovation, we delve into the multifaceted approach required to cultivate a mindset of curiosity, creativity, and problem-solving among young minds.

The narrative culminates in a vision of a developed India, a nation that has realized its potential across economic, technological, social, and cultural dimensions. Innovation, sustainability, inclusivity, and global leadership define this vision, setting the stage for a prosperous and harmonious future for the nation and its people.

As we embark on this exploration, I extend my gratitude to the contributors, innovators, and thought leaders who have lent their insights to this compilation. May this journey into the realm of Indigenous Technologies inspire, ignite, and contribute to the collective vision of a transformed India.

Forward into the Future!

Preface

In the symphony of progress, innovation emerges as the guiding melody that propels societies forward. As we stand at the threshold of a new era, this compilation seeks to unravel the intricacies of Indigenous Technologies and their profound impact on shaping the narrative of a Developed India.

Our journey begins by peering into the challenges faced by innovators in securing funding, navigating the intricate landscape of government grants, private investments, and venture capital. Through the lens of successful funding models, we envision the creation of investment ecosystems that provide robust support for indigenous technological ventures, laying the groundwork for economic prosperity.

Technological barriers often cast shadows on the path of innovation. This compilation unravels the tapestry of solutions that transcend infrastructural challenges, providing insights into overcoming obstacles and expanding access to cutting-edge tools. The narrative then weaves through the realms of artificial intelligence and biotechnology, exploring how Indigenous Technologies are poised to revolutionize industries and contribute to the nation's holistic development.

Communication forms the bridge between innovation and societal impact. We delve into the challenges of articulating the value of these technologies to the public, shedding light on strategies to bridge the awareness gap. The role of education and outreach programs takes center stage, envisioning a cultural shift that embraces and celebrates indigenous innovations.

The focus then shifts to the myriad opportunities that arise from the embrace of indigenous technologies. The narrative explores how these technologies can spur economic growth, create employment opportunities, and contribute to sustainable development. The chapter also delves into the potential for global collaborations and market expansion, presenting a vision of India as a global technological leader.

Emphasizing the need to cultivate a culture of innovation and self-reliance, we explore how educational institutions, research centers, and industries can collaboratively nurture a mindset that values indigenous technologies. The chapter concludes by leaving readers with a vision of a future where innovation becomes an integral part of the nation's DNA, propelling continuous progress.

The gaze then turns toward the future, envisioning the trajectory of indigenous technologies in shaping the destiny of India. The narrative unfolds into the promising landscape of emerging fields and the continued growth of traditional knowledge systems, offering a glimpse into a future driven by innovation and technological prowess.

As the integration of cutting-edge technologies becomes paramount, our exploration extends to the potential of indigenous technologies in emerging fields, with a special focus on artificial intelligence and biotechnology. From AI-driven solutions to genetic advancements, the chapter delves into how these technologies can revolutionize industries and contribute to the overall development of the nation.

The narrative takes a moment to emphasize the importance of preserving and promoting traditional knowledge systems. By tapping into the wealth of wisdom passed down through generations, India can harness the strengths of traditional practices in agriculture, healthcare, and more. The exploration unfolds the symbiotic relationship between ancient wisdom and contemporary innovation.

The compilation then explores how indigenous technologies can contribute to sustainable development, addressing environmental challenges, and fostering eco-friendly practices. From renewable energy solutions to waste management innovations, the chapter envisions a future where indigenous technologies play a pivotal role in achieving balance and harmony with nature.

Nurturing the next generation of innovators takes center stage in our exploration. The chapter delves into various aspects of fostering innovation in the next generation, including education, mentorship, and creating conducive environments for innovation.

The narrative concludes with an epilogue, painting a vision of a Developed India that has harnessed its potential across economic, technological, social, and cultural dimensions. Innovation, sustainability, inclusivity, and global leadership define this vision, setting the stage for a prosperous and harmonious future for the nation and its people.

As we embark on this journey into the realm of Indigenous Technologies, we extend our gratitude to the contributors, innovators, and thought leaders who have lent their insights to this compilation. May this exploration inspire, ignite, and contribute to the collective vision of a transformed India.

Unveiling the Tapestry of Indigenous Innovation!

Prologue

In the grand tapestry of a nation's evolution, innovation serves as the thread that weaves progress and resilience into the fabric of its identity. This compilation embarks on a journey, a prelude to the exploration of Indigenous Technologies and their transformative potential in sculpting the narrative of a Developed India.

As we step into the realms of funding challenges, the very cornerstone of innovation, we delve into the intricate dance between innovators and financial support. Government grants, private investments, and venture capital emerge as the pillars that uphold the dreams of those daring to pioneer Indigenous Technologies. This prologue sets the stage for a symphony of economic growth and technological advancement.

The narrative unfolds, revealing the technological barriers that stand as gatekeepers to progress. From infrastructural challenges to limited access to cutting-edge tools, we navigate through the solutions that pave the way for innovation to break free from its shackles. The spotlight then turns to artificial intelligence and biotechnology, where Indigenous Technologies emerge as torchbearers, illuminating the path toward a new era of advancement.

Communication becomes the heartbeat of progress, and we explore the challenges of conveying the value of technologies to the public. The prologue sets the scene for strategies to bridge the awareness gap, envisioning a future where education and outreach programs foster a cultural embrace of indigenous innovations.

The narrative then shifts to the opportunities that await, a myriad of possibilities stemming from the embrace of indigenous technologies. Economic growth, employment creation, and sustainable development take center stage, foreshadowing a future where India's technological prowess reaches global heights. The prologue emphasizes the potential for collaborations and market expansion, painting a vision of a nation at the forefront of innovation.

Emphasizing the need to cultivate a culture of innovation and self-reliance, we set the tone for collaboration between educational institutions, research centers, and industries. The prologue sparks the imagination for a future where innovation becomes an integral part of the nation's identity, propelling continuous progress.

As our gaze shifts toward the future, we catch a glimpse of the trajectory of indigenous technologies shaping the destiny of India. Emerging fields and the growth of traditional knowledge systems beckon, offering a vision of a nation driven by innovation and technological prowess.

The integration of cutting-edge technologies becomes the epicenter of our exploration, with a special focus on artificial intelligence and biotechnology. From AI-driven solutions to genetic advancements, the prologue delves into how these technologies can revolutionize industries and contribute to the nation's comprehensive development.

Preserving and promoting traditional knowledge systems come into focus, emphasizing the symbiotic relationship between ancient wisdom and contemporary innovation. The prologue sets the stage for tapping into the wealth of wisdom passed down through generations, harnessing the strengths of traditional practices in agriculture, healthcare, and more.

In the pursuit of sustainable development, the prologue envisions indigenous technologies addressing environmental challenges and fostering eco-friendly practices. From renewable energy solutions to waste management innovations, the compilation lays the groundwork for a future where indigenous technologies play a pivotal role in achieving balance and harmony with nature.

Nurturing the next generation of innovators takes center stage in our exploration, with the prologue setting the foundation for cultivating a mindset of curiosity, creativity, and problem-solving among young minds.

As we step into this exploration, we extend our gratitude to the contributors, innovators, and thought leaders who have shared their insights. May this journey into the realm of Indigenous Technologies serve as an inspiration, igniting the flame of innovation that propels India into a future of boundless possibilities.

A Prelude to Indigenous Ingenuity!

Challenges And Opportunities

7.

Future Prospects

Epilogue: A Vision of Developed India

CHAPTER ONE

Introduction to Indigenous Technologies

In the quest for a Developed India, the role of Indigenous Technologies emerges as a pivotal force driving innovation, economic growth, and technological self-sufficiency. This chapter serves as a compass guiding our exploration into the profound impact of indigenous technologies on the nation's development.

1.1 The Significance of Indigenous Technologies

The chapter opens with an examination of the fundamental importance of indigenous technologies. By exploring the historical context and understanding how traditional knowledge intersects with modern advancements, we lay the groundwork for comprehending the unique strengths that indigenous technologies bring to the table. The chapter delves into how these technologies bridge the gap between tradition and progress, fostering a dynamic ecosystem.

The examination of the fundamental importance of indigenous technologies involves a comprehensive exploration of their historical context and the intersection of traditional knowledge with modern advancements. This exploration is crucial for laying the groundwork to comprehend the unique strengths that indigenous technologies bring to the table. By delving into how these technologies bridge the gap between tradition and progress, a dynamic ecosystem is fostered.

Historical Context: Understanding the historical context of indigenous technologies is vital to appreciate their evolution and resilience over time. Indigenous communities have developed technologies that are deeply rooted in their cultural practices, passed down through generations. These technologies have often been sustainable, resource-efficient, and tailored to the specific needs of the community and its environment.

Intersecting Traditional Knowledge with Modern Advancements: The intersection of traditional knowledge with modern advancements is a key aspect of indigenous technologies. Many indigenous communities have demonstrated a remarkable ability to adapt traditional practices to incorporate modern tools and techniques. This blending of the old and the new not only preserves cultural heritage but also enhances the efficacy and relevance of indigenous technologies in the contemporary world.

Unique Strengths of Indigenous Technologies: Indigenous technologies offer unique strengths that arise from their intimate connection with local ecosystems and cultural values. These technologies are often characterized by their sustainability, adaptability, and community-centric approach. They leverage traditional wisdom to address modern challenges, providing innovative solutions that are culturally sensitive and environmentally sustainable.

Bridging the Gap Between Tradition and Progress: One of the most significant contributions of indigenous technologies is their role in bridging the gap between tradition and progress. Instead of viewing tradition and modernity as conflicting forces, indigenous technologies demonstrate how these two can coexist harmoniously. This integration fosters a holistic approach to development, ensuring that progress is achieved without sacrificing cultural identity or environmental balance.

Fostering a Dynamic Ecosystem: The exploration of indigenous technologies reveals that they contribute to the creation of a dynamic ecosystem. This ecosystem is characterized by a symbiotic relationship between traditional knowledge and contemporary advancements, promoting innovation, resilience, and inclusivity. Indigenous technologies become a driving force in sustainable development, encouraging a balance between technological progress and cultural preservation.

In conclusion, the examination of the fundamental importance of indigenous technologies provides valuable insights into their historical roots, the synergy between traditional knowledge and modern advancements, and the unique strengths they bring to the global technological landscape. Recognizing and appreciating these contributions not only enriches our understanding of diverse cultures but also inspires a more inclusive and sustainable approach to technological development.

1.2 Fostering Innovation, Research, and Development

To propel India towards becoming a global technological powerhouse, it is imperative to nurture an environment that fosters innovation, research, and development. This section explores the initiatives, policies, and institutions that play a pivotal role in encouraging scientific endeavors within the country. From academic research to industrial collaborations, the chapter highlights the multifaceted approach necessary for sustainable technological growth.

The examination of the fundamental importance of indigenous technologies is crucial for appreciating the unique contributions they bring to the global technological landscape. By delving into the historical context, we gain insights into how traditional knowledge has evolved and intersected with modern advancements, creating a rich tapestry of innovation. This exploration lays the groundwork for comprehending the unique strengths that indigenous technologies offer.

Indigenous technologies, rooted in the wisdom and practices of local communities, often represent a deep understanding of the environment, sustainable practices, and resource management. This knowledge, passed down through generations, provides a foundation for innovation that is not only culturally significant but also environmentally conscious.

In the context of technological progress, indigenous technologies serve as a bridge between tradition and modernity. They encapsulate the wisdom of the past while integrating with contemporary tools and methodologies. This synthesis fosters a dynamic ecosystem where traditional practices coexist and complement modern advancements, creating a harmonious balance.

Furthermore, the examination of indigenous technologies reveals their adaptability and resilience. These technologies often emerge from a context of limited resources, challenging environments, and specific cultural needs. This context-driven innovation results in solutions that are not only effective but also sustainable and contextually relevant.

One notable aspect is the holistic approach of indigenous technologies, where they address not only the technical aspects but also encompass cultural, social, and environmental dimensions. This holistic perspective is increasingly recognized as valuable in addressing complex global challenges, such as climate change, resource depletion, and sustainable development.

In conclusion, understanding the fundamental importance of indigenous technologies involves appreciating their historical context, recognizing the intersection of traditional knowledge with modern advancements, and acknowledging the unique strengths they bring to the global technological landscape. By valuing and integrating indigenous technologies, societies can tap into a rich source of innovation that promotes sustainability, cultural continuity, and a harmonious coexistence of tradition and progress.

1.3 Impact on Economic Growth and Self-Sustainability

Economic prosperity and self-sustainability are intertwined with the successful integration of indigenous technologies. By examining case studies and success stories, the chapter illustrates how leveraging homegrown innovations can enhance economic growth and reduce dependence on external sources. From grassroots initiatives to large-scale industries, we explore the far-reaching implications of indigenous technologies on various sectors of the economy.

The integration of indigenous technologies can play a crucial role in fostering economic prosperity and self-sustainability. Let's delve into some case studies and success stories that highlight the positive impact of leveraging homegrown innovations:

India's IT Industry:

Case Study: India's information technology (IT) industry is a prime example of successful integration of indigenous technologies. Starting from the 1980s, Indian companies like Tata Consultancy Services (TCS), Infosys, and Wipro have developed cutting-edge software solutions and services. This has not only led to economic growth but has also positioned India as a global IT hub.

Green Energy Solutions in Kenya:

Case Study: Kenya has embraced indigenous technologies to address its energy needs. The success story of M-KOPA, a company providing off-grid solar solutions, exemplifies this. M-KOPA has leveraged mobile technology to offer affordable solar power solutions to rural households, contributing to economic development while promoting sustainability.

China's High-Speed Rail Technology:

Case Study: China's development of high-speed rail technology is a testament to its commitment to indigenous innovation. The country has successfully implemented its own high-speed rail network, becoming a global leader in the industry. This has not only boosted the domestic economy but has also created opportunities for international collaborations and exports.

Brazil's Biofuel Industry:

Case Study: Brazil has achieved self-sustainability in the energy sector through the development of biofuels, particularly ethanol from sugarcane. By investing in indigenous technologies for biofuel production, Brazil has reduced its dependence on imported fossil fuels and created a thriving bioenergy industry, contributing to economic growth and environmental sustainability.

Rwanda's E-Government Initiatives:

Case Study: Rwanda has successfully implemented e-government initiatives to improve public services and governance. The use of indigenous technologies in areas such as digital payments, online services, and data management has not only enhanced efficiency but has also stimulated economic growth by creating a conducive environment for businesses and investment.

Indigenous Agricultural Technologies in Nigeria:

Case Study: In Nigeria, initiatives such as the Oyo State Agricultural Input Supply Company (OYSAIC) showcase the impact of indigenous agricultural technologies. OYSAIC focuses on developing and distributing locally adapted and cost-effective agricultural inputs, contributing to increased productivity, rural development, and economic resilience.

These case studies demonstrate that leveraging indigenous technologies can have far-reaching implications across various sectors, from information technology and energy to transportation, agriculture, and governance. The successful integration of homegrown innovations not only drives economic growth but also reduces dependency on external sources, fostering self-sustainability and resilience in the face of global challenges.

1.4 Technological Advancement: A National Imperative

The chapter underscores the urgency of technological advancement in shaping the future of a Developed India. By showcasing instances where indigenous technologies have propelled the nation forward, we establish a compelling narrative for the integration of these technologies into mainstream development strategies. It emphasizes how a commitment to innovation can position India at the forefront of the global technological landscape.

Absolutely, the urgency of technological advancement is crucial for shaping the future of a developed India. Let's highlight some instances where indigenous technologies have played a pivotal role in propelling the nation forward, establishing a compelling narrative for their integration into mainstream development strategies:

Digital India Campaign:

Instance: The Digital India campaign, launched by the Indian government, exemplifies the commitment to technological advancement. It focuses on transforming India into a digitally empowered society and knowledge economy. Initiatives like Aadhaar, UPI (Unified Payments Interface), and digital governance platforms showcase how indigenous technologies are driving efficiency, transparency, and inclusivity.

ISRO's Space Achievements:

Instance: The Indian Space Research Organisation (ISRO) has demonstrated significant prowess in space technology. The Mars Orbiter Mission (Mangalyaan) and the successful launch of a large number of satellites through cost-effective methods illustrate India's indigenous capabilities in space exploration. Such achievements not only contribute to scientific advancements but also enhance India's global standing in space technology.

Make in India in Defense:

Instance: The Make in India initiative, particularly in the defense sector, emphasizes the importance of indigenous technologies. Efforts to manufacture defense equipment, such as Tejas fighter jets, showcase how India is striving to reduce dependency on foreign imports and build a robust defense industry based on indigenous innovation and production.

BharatNet Project:

Instance: The BharatNet project, aiming to connect rural areas with high-speed internet, highlights the role of indigenous technologies in bridging the digital divide. By developing and implementing cost-effective solutions for internet connectivity, India can empower rural communities, spur economic development, and enhance educational opportunities.

Start-up Ecosystem:

Instance: India's vibrant start-up ecosystem is a testament to the entrepreneurial spirit and innovation in the country. Numerous start-ups across various sectors, from fintech to healthcare, are developing cutting-edge solutions. The success of these start-ups not only contributes to economic growth but also positions India as a hub for technological innovation on the global stage.

Renewable Energy Initiatives:

Instance: India's focus on renewable energy, with initiatives like the National Solar

Mission, emphasizes the integration of indigenous technologies for sustainable development. Advances in solar technology and the promotion of domestically produced renewable energy equipment showcase how innovation can drive environmental sustainability and economic growth simultaneously.

By showcasing these instances, we establish a compelling narrative for the integration of indigenous technologies into mainstream development strategies. It emphasizes how a commitment to innovation can position India at the forefront of the global technological landscape, fostering economic development, self-reliance, and competitiveness on the world stage. The urgency lies in leveraging these capabilities to address current challenges and shape a prosperous and technologically advanced future for the nation.

1.5 Setting the Stage for Exploration

As we conclude this introductory chapter, the stage is set for a comprehensive exploration of the rich history and potential of indigenous technologies in India. From the ancient roots of traditional knowledge systems to the cutting-edge advancements in contemporary science, the narrative unfolds to reveal the multifaceted journey towards a technologically empowered nation. The subsequent chapters will delve deeper into specific aspects, beginning with a tribute to the Indian science giants whose contributions have laid the foundation for the indigenous technological landscape.

In this introductory chapter, we embark on a journey to unravel the rich history and untapped potential of indigenous technologies in India. The narrative takes us from the ancient roots of traditional knowledge systems to the forefront of contemporary scientific advancements, setting the stage for a comprehensive exploration. The goal is to understand how this amalgamation of ancient wisdom and modern innovation contributes to shaping India as a technologically empowered nation.

The historical context is crucial as we trace the evolution of indigenous technologies, which have been deeply embedded in the cultural fabric of India. Traditional knowledge systems, passed down through generations, have not only survived the test of time but continue to play a pivotal role in shaping technological landscapes. This chapter aims to illuminate the symbiotic relationship between ancient practices and cutting-edge advancements, showcasing how India's technological journey is rooted in its diverse history.

As we progress, subsequent chapters will delve deeper into specific aspects of indigenous technologies. The journey begins with a tribute to the Indian science giants whose monumental contributions have laid the foundation for the indigenous technological landscape. By acknowledging their endeavors, we honor the pioneers who have paved the way for innovation and inspired future generations.

This exploration goes beyond a mere historical account; it aims to highlight the interconnectedness of tradition and progress, showcasing how indigenous technologies serve as a bridge between the old and the new. Through this lens, we aim to appreciate the unique strengths and perspectives that indigenous technologies bring to the table, fostering a holistic understanding of India's technological prowess.

As we navigate through the chapters, each facet of indigenous technologies will be scrutinized, shedding light on their significance in contemporary times. From sustainable practices rooted in ancient wisdom to the adaptability and resilience displayed in the face of modern challenges, we will unravel the multifaceted layers that make indigenous technologies a cornerstone of India's technological landscape. This journey invites readers to appreciate the dynamism, innovation, and potential embedded in the fusion of tradition and progress, ultimately contributing to the nation's advancement as a global technological powerhouse.

CHAPTER TWO

National Science Day

2.1 Commemorating Scientific Excellence

National Science Day, observed annually on February 28th, stands as a beacon of scientific achievement and a celebration of intellectual prowess. This chapter delves into the significance of this day, paying homage to the groundbreaking discovery of the Raman Effect by Sir C.V. Raman, and how it has evolved into a symbolic tribute to the immense contributions of Indian scientists.

National Science Day is indeed observed annually on February 28th in India to commemorate the discovery of the Raman Effect by Sir C.V. Raman in 1928. This discovery marked a significant milestone in the field of science and garnered international acclaim. The Raman Effect refers to the inelastic scattering of light by molecules, providing insights into molecular vibrations and allowing for a deeper understanding of the composition and structure of matter.

Sir C.V. Raman was awarded the Nobel Prize in Physics in 1930 for this groundbreaking discovery, making him the first Asian and the first non-white person to receive a Nobel Prize in any branch of science. His achievement brought global recognition to Indian science and laid the foundation for future scientific endeavors in the country.

The significance of National Science Day goes beyond just celebrating one individual's achievement. It serves as a platform to acknowledge and appreciate the broader contributions of the scientific community in India. The day is marked by various events, seminars, and educational activities aimed at promoting scientific temper and fostering a spirit of inquiry and innovation among the youth.

The theme for National Science Day is chosen each year to highlight specific areas of scientific development and research. This allows for a diverse range of topics to be explored and discussed, encouraging scientific curiosity among people of all ages. The day also emphasizes the importance of scientific research in addressing societal challenges and fostering national development.

National Science Day is not only a time to reflect on the achievements of the past but also a call to action for the future. It aims to inspire and motivate young minds to pursue careers in science, technology, engineering, and mathematics (STEM) fields, contributing to the overall progress and development of the nation.

In conclusion, National Science Day stands as a symbol of scientific excellence and intellectual prowess in India. It honors the legacy of Sir C.V. Raman and serves as a reminder of the transformative power of scientific discovery. The day encourages a culture of scientific inquiry, innovation, and education, paving the way for a brighter and more technologically advanced future for the country.

2.2 The Raman Effect: Pioneering Scientific Achievement

At the heart of National Science Day lies the discovery of the Raman Effect in 1928, a scientific breakthrough that earned Sir C.V. Raman the Nobel Prize in Physics. This section explores the intricacies of this discovery, shedding light on its impact not only on the global scientific community but also on the trajectory of scientific research in India. We delve into the scientific ingenuity behind the Raman Effect and its enduring legacy.

Absolutely, National Science Day in India commemorates the discovery of the Raman Effect on February 28, 1928, by the renowned physicist Sir C.V. Raman. The Raman Effect refers to the inelastic scattering of light by matter, and it played a pivotal role in the understanding of molecular and atomic interactions.

Here's a brief overview of the significance of the Raman Effect and its impact:

Discovery of the Raman Effect:

In 1928, while working at the Indian Association for the Cultivation of Science in Kolkata, Sir C.V. Raman discovered that when light interacts with molecules, it scatters and undergoes a change in wavelength. This phenomenon, known as the Raman Effect, provided valuable insights into the vibrational and rotational modes of molecules.

Scientific Ingenuity:

Raman's discovery showcased his exceptional scientific insight and experimental skills. His ability to observe subtle changes in light scattering allowed him to uncover a fundamental aspect of molecular behavior.

Nobel Prize in Physics (1930):

The significance of Raman's work was recognized globally, leading to him being awarded the Nobel Prize in Physics in 1930. He became the first Asian and the first non-white individual to receive a Nobel Prize in the sciences.

Impact on Global Scientific Community:

The Raman Effect opened up new avenues for the study of molecular structure and dynamics. It became a crucial tool in various scientific disciplines, including chemistry, physics, biology, and material science.

Legacy in India:

Sir C.V. Raman's discovery had a profound impact on the trajectory of scientific research in India. It inspired a generation of scientists and contributed to the country's reputation in the field of science.

Technological Applications:

The Raman Effect has found numerous applications in diverse fields. Raman spectroscopy, based on this effect, is widely used in chemical analysis, material science, pharmaceuticals, and biomedical research.

Educational Initiatives:

Raman's legacy has been instrumental in fostering scientific education and research in India. National Science Day serves as a reminder of the importance of scientific inquiry and the role of Indian scientists in contributing to global knowledge.

In summary, the Raman Effect stands as a testament to the power of scientific curiosity and the impact that a single discovery can have on the scientific community, both locally and globally. National Science Day provides an opportunity to celebrate and reflect on such milestones in the history of science.

2.3 Themes and Activities: Nurturing Scientific Temper

National Science Day serves as a platform for a diverse array of themes and activities that engage people from all walks of life. The chapter investigates the annual themes chosen to align with contemporary scientific challenges and opportunities. From symposiums and exhibitions to educational programs and public lectures, the day unfolds as a tapestry of events designed to foster scientific temper and curiosity.

National Science Day in India is not just a commemoration of a historical scientific breakthrough but also an occasion to promote scientific awareness, temper, and curiosity across the nation. Each year, a specific theme is chosen to align with contemporary scientific challenges and opportunities. This theme serves as a focal point for various activities and events organized to engage people from all walks of life. Here's an exploration of the diverse array of themes and activities associated with National Science Day:

Annual Themes:

The annual themes are carefully chosen to reflect current scientific issues and advancements. These themes can range from specific scientific topics to broader societal challenges that science can help address.

Symposiums and Exhibitions:

National Science Day is marked by symposiums and exhibitions where scientists, researchers, and experts share their insights and discoveries. These events provide a platform for the exchange of knowledge and ideas.

Educational Programs:

Educational initiatives play a significant role on National Science Day. Workshops, seminars, and educational programs are organized to inspire and educate students about the wonders of science, encouraging them to pursue careers in scientific research.

Public Lectures:

Renowned scientists often deliver public lectures on National Science Day, making science accessible to a wider audience. These lectures may cover a range of topics, from fundamental scientific principles to cutting-edge research.

Science Exhibitions for Students:

Exhibitions designed for students are a common feature, showcasing scientific experiments, models, and innovations. This hands-on approach helps in making science more tangible and interesting for young minds.

Science Competitions:

National Science Day often sees the organization of science competitions for school and college students. These competitions may include quizzes, debates, and science fairs, fostering healthy competition and a spirit of inquiry.

Science Outreach Programs:

To reach a broader audience, science outreach programs are organized, taking science to communities that may have limited access to scientific resources. This inclusivity is crucial for promoting scientific temper across diverse demographics.

Collaboration with Institutions:

National Science Day encourages collaboration between scientific institutions, universities, and research organizations. Joint efforts amplify the impact of the celebrations and create a sense of unity in the scientific community.

Media Campaigns:

The day is often accompanied by media campaigns to raise public awareness about the importance of science and its impact on society. This can include articles, documentaries, and interviews with scientists.

In essence, National Science Day is a multifaceted celebration that goes beyond a historical event. It serves as a catalyst for instilling a scientific temperament in the public, inspiring the younger generation, and fostering a culture of curiosity and inquiry across the country. The diverse range of activities ensures that people from all backgrounds can participate and appreciate the significance of science in our lives.

2.4 Promoting Scientific Temper Among the Masses

Beyond the ceremonial aspects, the chapter explores how National Science Day plays a crucial role in promoting scientific temper among the masses. It examines educational initiatives, outreach programs, and the dissemination of scientific knowledge to schools and communities. The day becomes a catalyst for inspiring the next generation of scientists, instilling a sense of curiosity and wonder about the natural world.

National Science Day plays a crucial role in promoting scientific temper and fostering a scientific mindset among the masses. Beyond the ceremonial aspects, the day is marked by various educational initiatives, outreach programs, and the dissemination of scientific knowledge to schools and communities.

Educational Initiatives:

School Programs: Many schools organize special science-related activities, workshops, and exhibitions to engage students in hands-on experiments and demonstrations. These initiatives aim to make science more accessible and interesting to students, encouraging them to develop a keen interest in the subject.

Science Competitions: National Science Day often sees the organization of science competitions at various levels. These competitions provide a platform for students to showcase their scientific knowledge and innovative ideas, fostering healthy competition and enthusiasm for scientific inquiry.

Outreach Programs:

Science Fairs and Exhibitions: Science fairs and exhibitions are organized to showcase scientific advancements, experiments, and innovations. These events are not only informative but also interactive, allowing the public to engage with scientists and researchers and understand the practical applications of scientific concepts.

Public Lectures and Talks: Eminent scientists and researchers often deliver public lectures and talks to explain complex scientific ideas in a comprehensible manner. These events aim to bridge the gap between the scientific community and the general public, fostering a better understanding of scientific principles.

Dissemination of Scientific Knowledge:

Media and Communication: National Science Day is an opportunity to leverage various media platforms to disseminate scientific knowledge. Television programs, radio shows, and articles in newspapers and magazines can be dedicated to popularizing scientific achievements, discoveries, and their impact on society.

Online Platforms: With the advent of technology, online platforms, including social media, websites, and webinars, play a crucial role in reaching a wider audience. National Science Day can be used as a launchpad for online campaigns and initiatives to promote scientific literacy and awareness.

Inspiring the Next Generation:

Role Models and Mentorship: Highlighting the contributions of eminent scientists, including Sir C.V. Raman, serves as inspiration for young minds. The day can be used to celebrate the achievements of contemporary scientists, creating role models who can inspire the next generation to pursue careers in science and research.

Hands-on Learning: Practical and hands-on learning experiences, such as laboratory visits and experiments, can ignite curiosity and a passion for scientific exploration. National Science Day provides a platform to organize such activities, making science more tangible and exciting for students.

In essence, National Science Day becomes a catalyst for inspiring the next generation of scientists. Through educational initiatives, outreach programs, and the dissemination of scientific knowledge, the day contributes to the development of a scientifically informed and curious society. It instills a sense of wonder about the natural world and encourages individuals, especially the youth, to explore the realms of science and contribute to the progress of society.

2.5 Indian Scientists on the Global Stage

National Science Day is also an opportunity to reflect on the broader contributions of Indian scientists to the global scientific community. By showcasing the achievements and recognition garnered by Indian scientists on an international scale, the chapter underscores the importance of India's role in the advancement of global scientific knowledge.

Absolutely, National Science Day serves as a platform to reflect on the significant contributions of Indian scientists to the global scientific community. By showcasing their achievements and international recognition, the day underscores the importance of India's role in the advancement of global scientific knowledge. Here are some key points that the chapter could explore:

Global Recognition of Indian Scientists:

Highlighting instances where Indian scientists have received international accolades, awards, and honors for their contributions to various scientific fields. This includes Nobel

Prizes, prestigious fellowships, and leadership roles in international scientific organizations.

Pioneering Research and Discoveries:

Showcasing specific instances where Indian scientists have been at the forefront of pioneering research and groundbreaking discoveries that have had a global impact. This could include advancements in space exploration, medical research, information technology, and other critical scientific domains.

Leadership in International Collaborations:

Discussing India's active participation in international scientific collaborations and partnerships. This could involve joint research projects, collaborative experiments, and contributions to global scientific initiatives, showcasing the country's commitment to fostering a collaborative and inclusive scientific community.

Scientific Diaspora:

Recognizing the achievements of the Indian scientific diaspora—scientists of Indian origin working abroad who have made substantial contributions to their respective fields. This highlights the global reach and influence of Indian scientists in various scientific disciplines.

Technological Innovation and Impact:

Exploring instances where Indian scientists and researchers have played a pivotal role in technological innovation with global implications. This could include advancements in areas such as artificial intelligence, renewable energy, and biotechnology.

Scientific Institutions and Infrastructure:

Highlighting the growth and development of scientific institutions and infrastructure in India that contribute to the global scientific landscape. This includes research facilities, laboratories, and academic institutions that foster innovation and collaboration.

Challenges and Opportunities:

Acknowledging the challenges faced by Indian scientists, such as funding constraints and infrastructure limitations, while also emphasizing the opportunities for growth and expansion in the global scientific arena.

By examining the broader contributions of Indian scientists on the international stage, the chapter can convey the message that National Science Day is not only a celebration of past achievements but also a recognition of India's ongoing commitment to advancing global scientific knowledge and addressing global challenges through scientific inquiry and innovation.

2.6 Future Aspirations: Building on the Legacy

As we conclude this chapter, the focus shifts towards the future. We explore the aspirations and potential avenues for National Science Day to continue serving as a catalyst for scientific exploration, innovation, and collaboration. By building on the legacy of Sir C.V. Raman and other Indian science giants, the day becomes a stepping stone towards a future where scientific excellence is celebrated, nurtured, and ingrained in the cultural fabric of the nation. The subsequent chapters will further unravel the tapestry of Indigenous Technologies, building upon the foundation laid by the commemorative spirit of National Science Day.

The future of National Science Day holds great promise as a catalyst for scientific exploration and innovation. By continuing to build on the legacy of Sir C.V. Raman and other influential Indian scientists, the day can serve as a stepping stone towards a future where scientific excellence is not only celebrated but also actively nurtured. Here are some potential avenues for National Science Day to continue playing a pivotal role in shaping the scientific landscape:

Promoting Scientific Literacy:

National Science Day can focus on enhancing scientific literacy among the general public. Initiatives such as science fairs, workshops, and interactive exhibitions can be organized to make science more accessible and engaging.

Youth Engagement and Education:

Emphasizing educational programs that target young minds can inspire the next generation of scientists. Schools, colleges, and universities can collaborate on events like science competitions, hackathons, and career guidance sessions.

Showcasing Indigenous Technologies:

Future National Science Days could spotlight indigenous technologies that blend traditional knowledge with modern scientific advancements. This could foster a deeper appreciation for India's rich scientific heritage and its potential to address contemporary challenges.

International Collaboration:

Facilitating collaborations with international scientific communities can open doors for knowledge exchange and collaborative research. This can contribute to India's standing in the global scientific arena.

Inclusive Participation:

Ensuring the active involvement of scientists and researchers from diverse backgrounds, including women and marginalized communities, can promote inclusivity and diversity in scientific pursuits.

Harnessing Technological Platforms:

Leveraging technology for virtual symposiums, online exhibitions, and interactive webinars can broaden the reach of National Science Day, allowing people from various regions to participate and benefit.

Supporting Innovation Ecosystems:

National Science Day can contribute to the growth of innovation ecosystems by supporting startup initiatives, providing grants for research projects, and fostering collaborations between academia and industry.

Environmental and Sustainability Focus:

Aligning National Science Day themes with pressing global challenges, such as climate change and sustainability, can encourage scientific endeavors that address critical issues facing society.

Public-Private Partnerships:

Encouraging partnerships between the government, private sector, and non-profit organizations can create a synergistic environment for scientific research, development, and application.

By weaving these elements into the fabric of National Science Day, the event can continue to evolve as a dynamic platform that not only commemorates past achievements but also propels India towards a future where scientific curiosity and innovation are integral to the national ethos. The subsequent chapters exploring Indigenous Technologies will further contribute to this vision, highlighting how traditional knowledge and modern science can converge for the benefit of society.

CHAPTER THREE

Indian Science Giants

3.1 Introduction to Pioneers

This chapter unfolds as a tribute to the luminaries who have etched their names in the annals of science, contributing significantly to the global scientific community. From the early 20th century to the modern era, Indian scientists have propelled the nation into the forefront of scientific discovery, making invaluable contributions that resonate across disciplines.

Indeed, the scientific journey of India has been marked by the remarkable achievements of numerous luminaries who have left an indelible mark on the global scientific landscape. The saga begins in the early 20th century, where pioneering Indian scientists laid the foundation for future generations.

One such luminary is Sir Jagadish Chandra Bose, a polymath whose work spanned physics, biology, and archaeology. In the early 20th century, he demonstrated the similarity between animal and plant tissues, laying the groundwork for the modern study of bioelectricity and biophysics.

Moving forward, Dr. C.V. Raman's groundbreaking research in the 1920s on the scattering of light earned him the Nobel Prize in Physics in 1930. His discoveries, known as the Raman Effect, have had far-reaching applications in various scientific fields.

The mid-20th century witnessed the contributions of Dr. Homi J. Bhabha, often regarded as the father of the Indian nuclear program. His vision and leadership played a pivotal role in establishing the Tata Institute of Fundamental Research (TIFR) and later, the Atomic Energy Commission of India.

The space age brought forth the brilliance of Dr. Vikram Sarabhai, the founding father of the Indian Space Research Organisation (ISRO). His foresight and determination led to India's first satellite, Aryabhata, and laid the foundation for the country's successful space endeavors.

In the realm of theoretical physics, Dr. Satyendra Nath Bose's collaboration with Albert Einstein resulted in the development of Bose-Einstein statistics and the theory of the Bose-Einstein condensate, contributing significantly to quantum mechanics.

Transitioning into the modern era, the achievements of Dr. A.P.J. Abdul Kalam, known as the "People's President" and the "Missile Man of India," are noteworthy. His leadership in the development of ballistic missile technology exemplifies India's prowess in defense research and technology.

Today, Indian scientists continue to make strides in various fields, including space exploration, information technology, medicine, and environmental science. Their contributions not only enrich the global scientific community but also serve as an inspiration for future generations of scientists, fostering a culture of innovation and discovery. The legacy of these luminaries reflects the spirit of inquiry and the commitment to pushing the boundaries of knowledge that defines the scientific journey of India.

3.2 Sir Jagadish Chandra Bose: The Pioneer of Radio Waves

The narrative begins with Sir Jagadish Chandra Bose, a polymath whose work in the late 19th and early 20th centuries laid the foundation for our understanding of radio waves. This section explores Bose's groundbreaking experiments, his instrumental role in the development of wireless communication, and his enduring legacy in modern telecommunications.

Sir Jagadish Chandra Bose, a polymath of the late 19th and early 20th centuries, stands as a pivotal figure in the history of science and technology, particularly in the realm of radio waves and wireless communication. Born in 1858 in British India, Bose's multifaceted contributions spanned physics, biology, and telecommunications.

Bose's groundbreaking experiments in the late 19th century played a crucial role in unraveling the mysteries of radio waves. In 1895, he demonstrated the transmission of electromagnetic waves over a short distance, predating Guglielmo Marconi's more well-known experiments. His research laid the foundation for the development of wireless communication, a technology that would revolutionize global connectivity in the years to come.

One of Bose's notable achievements was the creation of the "mercury coherer," a device that could detect radio waves. This innovation significantly improved the efficiency of radio wave reception and paved the way for the development of more advanced communication systems. Bose's work was not solely limited to scientific exploration; he also patented his inventions, emphasizing their practical applications.

Bose's instrumental role in the early days of wireless communication cannot be overstated. His pioneering experiments and inventions set the stage for further advancements in telecommunications. While Marconi is often credited with the invention of the radio, Bose's contributions have gained recognition for their significant impact on the field.

The legacy of Sir Jagadish Chandra Bose endures in modern telecommunications. His work laid the groundwork for the wireless technologies that form the backbone of our interconnected world today. Beyond his scientific achievements, Bose's commitment to innovation and his interdisciplinary approach serve as an inspiration for future generations of scientists and engineers.

As we delve into the narrative of Sir Jagadish Chandra Bose's life and work, we uncover not only the scientific milestones that shaped his era but also the enduring influence that continues to shape the landscape of telecommunications in the 21st century.

3.3 Dr. Homi Bhabha: Architect of India's Nuclear Program

Moving forward in time, the chapter delves into the life and contributions of Dr. Homi Bhabha, often hailed as the architect of India's nuclear program. This section explores Bhabha's pivotal role in establishing India as a nuclear power, his leadership in scientific institutions, and the far-reaching implications of his vision for the nation's scientific advancement.

Dr. Homi Bhabha, a towering figure in the realm of science and technology, is often celebrated as the architect of India's nuclear program. Born in 1909, Bhabha played a pivotal role in shaping the scientific landscape of post-independence India and establishing the country as a nuclear power.

Bhabha's journey began with his academic pursuits in physics, earning a doctorate from the University of Cambridge in 1935. His early research contributions were recognized globally, setting the stage for a remarkable career in science. In the post-independence era, Bhabha's vision extended beyond academic achievements, as he recognized the strategic importance of nuclear technology for India's development.

Under Bhabha's leadership, the Atomic Energy Commission (AEC) was established in 1948, and he became its first chairman. His foresight and determination laid the foundation for India's nuclear program, with the establishment of the Tata Institute of Fundamental Research (TIFR) in 1945, and the Atomic Energy Establishment Trombay

(AEET) in 1954, which later became the Bhabha Atomic Research Centre (BARC).

Bhabha's emphasis on indigenous scientific capabilities and self-reliance set the tone for India's nuclear ambitions. Despite international skepticism and limited resources, he championed the development of nuclear reactors, leading to India's first nuclear test, codenamed "Smiling Buddha," in 1974. This event marked India's entry into the league of nuclear-armed nations.

Beyond his contributions to the nuclear program, Bhabha's leadership extended to various scientific institutions. He played a crucial role in establishing the Indian National Committee for Space Research (INCOSPAR), laying the groundwork for India's space exploration endeavors. Bhabha's influence extended to education as well, with his efforts in promoting scientific research and education leaving an indelible mark.

The far-reaching implications of Dr. Homi Bhabha's vision are evident in India's scientific and technological advancements. His commitment to harnessing nuclear energy for peaceful purposes, coupled with a vision for self-reliance, has shaped India's trajectory in the field. The legacy of Bhabha extends beyond his scientific achievements to his role as a visionary leader who laid the groundwork for India's scientific and technological prowess.

Dr. Homi Bhabha's legacy extends not only to India's nuclear and space programs but also to his efforts in fostering scientific research and education. Recognizing the importance of nurturing young scientific talent, he played a pivotal role in the establishment of the Tata Institute of Fundamental Research (TIFR) in Mumbai. TIFR became a hub for cutting-edge research in various scientific disciplines, attracting bright minds and contributing significantly to India's scientific community.

Under Bhabha's guidance, TIFR became a center for fundamental research in physics, chemistry, biology, and mathematics. The institute's interdisciplinary approach and commitment to excellence laid the groundwork for advancements in various scientific fields. Bhabha's influence on the institute ensured a culture of innovation, collaboration, and a pursuit of knowledge that transcended traditional boundaries.

Beyond his role at TIFR, Bhabha's impact on Indian science is reflected in his efforts to establish the Tata Institute of Fundamental Research School of Mathematics (TIFR-SM) and the Tata Institute of Social Sciences (TISS). These institutions played crucial roles in shaping the academic landscape of India, fostering research and education in mathematics and social sciences, respectively.

Dr. Bhabha's vision for self-reliance and indigenous development also extended to the field of nuclear power. He advocated for the use of nuclear energy for peaceful purposes, emphasizing its potential in addressing India's energy needs. Bhabha's commitment to building scientific capabilities within the country contributed to the successful development of indigenous nuclear reactors and technologies.

The narrative further explores Bhabha's diplomatic efforts to navigate India's nuclear ambitions on the global stage. His leadership in international forums and negotiations showcased a commitment to responsible nuclear practices and peaceful coexistence. Bhabha's contributions paved the way for India to gain global recognition as a responsible nuclear power.

3.4 Dr. A.P.J. Abdul Kalam: The People's President and Space Pioneer

The narrative then shifts to the "People's President" and eminent aerospace engineer, Dr. A.P.J. Abdul Kalam. This section delves into Kalam's exceptional contributions to space exploration, including his leadership in the development of India's first indigenous satellite launch vehicle. The chapter celebrates Kalam's indomitable spirit and his tireless efforts to bring space technology to the forefront of India's scientific landscape.

Dr. A.P.J. Abdul Kalam, often referred to as the "People's President" and the "Missile Man of India," stands as a beacon of inspiration in the annals of Indian science and space exploration. A distinguished aerospace engineer, Dr. Kalam played a pivotal role in shaping India's capabilities in space technology, leaving an enduring legacy that resonates to this day.

In the 1970s and 1980s, Dr. Kalam spearheaded the Integrated Guided Missile Development Program (IGMDP), which led to the successful development of a series of missiles, including the Agni and Prithvi missiles. His expertise and leadership in missile technology showcased India's prowess in defense research and positioned the nation as a self-reliant player in strategic capabilities.

Dr. Kalam's journey in space exploration took a significant turn when he assumed the role of the Scientific Advisor to the Defense Minister in 1992. Subsequently, he became the Chief Scientific Advisor to the Prime Minister and later served as the President of India from 2002 to 2007.

One of the crowning achievements during Dr. Kalam's tenure was the successful development and launch of India's first indigenous satellite launch vehicle, the Polar Satellite Launch Vehicle (PSLV). Under his guidance, India entered the elite club of nations capable of launching satellites into space. The PSLV went on to become a reliable workhorse, facilitating numerous successful missions and placing India on the global map in space exploration.

Dr. Kalam's indomitable spirit, visionary leadership, and tireless efforts were instrumental in bringing space technology to the forefront of India's scientific landscape. His passion for harnessing technology for the betterment of society extended beyond national borders and resonated with a global audience.

Beyond his technical contributions, Dr. Kalam was a charismatic leader who inspired countless individuals, especially the youth, to dream big and strive for excellence. His vision for India included harnessing space technology for societal development, emphasizing the peaceful use of outer space for purposes such as telecommunication, weather forecasting, and resource mapping.

In celebrating Dr. Kalam's contributions to space exploration, we not only acknowledge the scientific advancements but also honor the spirit of innovation, perseverance, and national pride that he instilled in the scientific community and the people of India. Dr. A.P.J. Abdul Kalam's legacy continues to inspire generations to reach for the stars, both metaphorically and literally, in the pursuit of knowledge and progress.

Dr. Kalam's impact on space exploration extended beyond the technical realm; he advocated for the democratization of space technology and its applications for societal development. His vision encompassed using space assets for education, healthcare, agriculture, and disaster management. Dr. Kalam firmly believed that space technology could be a powerful tool to address the socio-economic challenges facing the nation.

Under his leadership, ISRO initiated missions that aimed at leveraging space technology for the benefit of the common person. The Indian National Satellite System (INSAT) and the Indian Remote Sensing (IRS) satellites were instrumental in revolutionizing communication, broadcasting, weather forecasting, and natural resource management. These applications not only improved the quality of life for millions but also showcased India's ability to utilize space technology for inclusive development.

One of the hallmark achievements during Dr. Kalam's tenure was the successful execution of the Chandrayaan-1 mission in 2008. India's first lunar probe, Chandrayaan-1, made significant discoveries, including the presence of water molecules on the moon's surface. This mission marked India's entry into lunar exploration and reinforced the country's space capabilities on the global stage.

Dr. Kalam's influence extended to nurturing and inspiring the next generation of scientists and engineers. He often interacted with students and young professionals, encouraging them to pursue careers in science and technology. His famous line, "Dream, dream, dream. Dreams transform into thoughts, and thoughts result in action," continues to resonate with aspiring scientists, emphasizing the importance of ambition and perseverance.

Even after his presidency, Dr. Kalam remained an advocate for space exploration and technological innovation. He continued to engage with educational institutions, contributing to the development of scientific temper and research culture among the youth.

3.5 Dr. Vikram Sarabhai: The Visionary of Indian Space Program

The exploration continues with Dr. Vikram Sarabhai, another visionary in the realm of space science. As the founder of the Indian Space Research Organisation (ISRO), Sarabhai's contributions in the 1960s paved the way for India's remarkable achievements in space exploration. This section sheds light on Sarabhai's foresight, his dedication to scientific education, and his instrumental role in establishing a robust space program for the nation.

Dr. Vikram Sarabhai, widely regarded as the father of the Indian space program, was born on August 12, 1919, in Ahmedabad, Gujarat, India. His life story is a remarkable journey of vision, leadership, and dedication to advancing science and technology in India.

Early Life and Education: Vikram Sarabhai hailed from a prominent industrialist family in India. His father, Ambalal Sarabhai, was an industrialist, and his mother, Sarla Devi, was a progressive social worker. Sarabhai showed an early interest in science, and his education took him to Cambridge University in the United Kingdom, where he studied natural sciences.

Return to India: After completing his studies abroad, Vikram Sarabhai returned to India with a vision to contribute to the scientific and technological progress of the nation. His diverse interests led him to make significant contributions not only in space science but also in fields such as nuclear physics, cosmic rays, and even the establishment of key institutions.

Establishment of Physical Research Laboratory (PRL): In 1947, at the young age of 28, Sarabhai founded the Physical

Research Laboratory (PRL) in Ahmedabad. PRL became the cradle of space sciences in India, focusing on cosmic rays and space research.

Founding the Indian National Committee for Space Research (INCOSPAR): In 1962, Sarabhai successfully convinced the Indian government to set up the Indian National Committee for Space Research (INCOSPAR), laying the foundation for India's space program. This committee eventually evolved into the Indian Space Research Organisation (ISRO).

Aryabhata Satellite and Space Research: Sarabhai's vision for India included harnessing space technology for peaceful purposes. In 1975, India launched its first satellite, Aryabhata, marking the country's entry into the field of space research.

Establishment of ISRO: Sarabhai's dream of a national space agency came to fruition with the establishment of the Indian Space Research Organisation (ISRO) in 1969. His unwavering commitment and leadership paved the way for India's space exploration endeavors.

Thumba Equatorial Rocket Launching Station: Sarabhai played a crucial role in setting up the Thumba Equatorial Rocket Launching Station (TERLS) in Thiruvananthapuram, which became a significant center for sounding rocket research.

Personal and Professional Recognition: Vikram Sarabhai received numerous awards and honors for his contributions to science and technology. He was elected as the President of the International Astronautical Federation (IAF) in 1962.

Legacy: Dr. Vikram Sarabhai's legacy extends beyond his technical achievements. He was a visionary leader who emphasized the importance of using science and technology for the benefit of society. His contributions laid the groundwork for India's subsequent successes in space exploration.

Tragically Short Life: Despite his profound impact, Dr. Vikram Sarabhai's life was tragically cut short when he passed away on December 30, 1971, at the age of 52. His vision and contributions, however, continue to inspire generations of scientists and engineers in India and beyond.

Multifaceted Contributions: Beyond space and scientific endeavors, Vikram Sarabhai was involved in various other fields. He played a pivotal role in establishing the Indian Institute of Management (IIM) Ahmedabad in collaboration with other notable personalities. His vision was not limited to space and science but extended to holistic development and education.

International Collaboration: Sarabhai understood the importance of international collaboration in advancing space research. He actively collaborated with NASA and other international space agencies, fostering partnerships that benefited India's space program.

Educational Initiatives: Sarabhai was a strong advocate for education and believed in nurturing young minds. He emphasized the need for educational institutions to focus on research and innovation. His efforts led to the establishment of several institutions, contributing to the growth of science and technology education in India.

Vision for Application of Space Technology: Dr. Sarabhai's vision for space technology went beyond exploration. He envisioned the application of space technology for communication, weather forecasting, resource mapping, and agricultural monitoring. Many of these applications have become integral to India's development.

Social Relevance: Sarabhai's commitment to the social relevance of scientific advancements was evident in his emphasis on using space technology for societal development. He believed that technology should not only serve the elite but should also address the needs of the common people, contributing to the overall progress of the nation.

Personal Characteristics: Vikram Sarabhai was known for his charismatic personality, humility, and a rare combination of scientific acumen and leadership skills. His ability to inspire and motivate people, coupled with a forward-looking approach, played a crucial role in garnering support for India's space program.

Continuing Legacy: The institutions and initiatives established by Sarabhai continue to thrive and contribute to India's scientific and technological landscape. His legacy is carried forward by the individuals he inspired, including prominent scientists and leaders in various fields.

Posthumous Recognition: Even after his passing, Vikram Sarabhai received posthumous honors, including the Padma Bhushan, one of India's highest civilian awards. His contributions have been commemorated through various awards, institutions, and events dedicated to his memory.

Dr. Vikram Sarabhai's life story is a testament to his passion for science, commitment to societal progress, and far-reaching vision. His pioneering efforts laid the foundation for India's space journey and continue to shape the country's scientific pursuits, making him an enduring icon in the history of Indian science and technology.

3.6 Dr. C.N.R. Rao: A Titan in Materials Science

The chapter concludes with the contributions of Dr. C.N.R. Rao, a prolific scientist in the field of materials science and chemistry. Rao's work has garnered international acclaim, and this section explores his groundbreaking research, his role in shaping scientific institutions, and his advocacy for scientific education and research in India.

Dr. Chintamani Nagesa Ramachandra Rao, commonly known as C.N.R. Rao, is an eminent Indian scientist and one of the world's leading solid-state and structural chemists. Born on June 30, 1934, in Bangalore (now Bengaluru), India, his life story is characterized by exceptional achievements, scholarly contributions, and leadership in scientific research and education.

Here is an overview of Dr. C.N.R. Rao's life and contributions:

Early Life and Education:

Dr. C.N.R. Rao's interest in science and chemistry began at a young age. Despite facing financial challenges, his dedication to education led him to pursue a Bachelor's degree in Science at Mysore University.

He went on to earn a Master's degree in Chemistry from the Banaras Hindu University, where he was awarded the first position in the university.

His academic journey continued at Purdue University in the United States, where he completed his Ph.D. in 1958 under the guidance of the renowned chemist Professor Hugh Marshall.

Academic Career and Research:

After completing his Ph.D., Rao returned to India and began his academic career at the Indian Institute of Science (IISc) in Bangalore.

His research contributions in the field of solid-state and structural chemistry earned him global recognition. He conducted groundbreaking work on transition metal oxides and contributed to the understanding of their properties.

Rao has published over 1,600 research papers and several books, making him one of the most prolific and cited chemists in the world.

Leadership Roles:

Dr. C.N.R. Rao held several leadership positions in Indian scientific institutions. He served as the Director of the Indian Institute of Science, Bangalore, and later as the

Chairman of the Scientific Advisory Council to the Prime Minister of India.

His leadership extended internationally, where he served as the Linus Pauling Research Professor at the University of California, Santa Barbara, and held various advisory roles.

Awards and Honors:

Dr. Rao has received numerous awards and honors for his outstanding contributions to science. Notably, he was honored with the Bharat Ratna, India's highest civilian award, in 2014.

He has been elected as a Fellow of the Royal Society and is a member of several prestigious academies and scientific organizations worldwide.

Legacy:

Dr. C.N.R. Rao's legacy is marked by his commitment to scientific research, dedication to education, and advocacy for the advancement of science and technology in India.

He has been an influential figure in shaping national science policies and fostering a culture of scientific inquiry and innovation.

Dr. C.N.R. Rao's life story serves as an inspiration to aspiring scientists and underscores the importance of dedication, perseverance, and a passion for knowledge in the pursuit of scientific excellence.

Scientific Contributions:

Dr. C.N.R. Rao's contributions to chemistry and materials science include significant advancements in the understanding of crystal structures, nanomaterials, and the properties of various compounds.

His work on two-dimensional materials, particularly graphene, has been groundbreaking. He collaborated with Andre Geim and Konstantin Novoselov, who later received the Nobel Prize in Physics for their work on graphene, establishing the material's unique properties.

Rao's research also extends to high-temperature superconductors, semiconductor nanoparticles, and catalysis, demonstrating the breadth and depth of his scientific inquiry.

Educational Initiatives:

Throughout his career, Rao has been dedicated to nurturing young scientific talent. He played a key role in establishing the Jawaharlal Nehru Centre for Advanced Scientific Research (JNCASR) in Bangalore, providing a platform for interdisciplinary research.

As a teacher and mentor, he has influenced generations of students and researchers. His commitment to education and fostering a spirit of inquiry has left an indelible mark on the academic landscape in India.

Advocacy for Science:

Dr. Rao has been an outspoken advocate for the advancement of science and technology in

India. He has consistently stressed the importance of investing in research and development to propel the nation towards scientific excellence.

His involvement in science policy, both as an advisor to the government and through various scientific bodies, reflects his commitment to shaping the trajectory of scientific research and education in the country.

Personal Achievements and Recognition:

In addition to the Bharat Ratna, Dr. C.N.R. Rao has received numerous international awards, including the Dan David Prize and the August Wilhelm von Hofmann Medal.

He has been associated with several scientific journals, including serving as the Editor-in-Chief of the Journal of Solid State Chemistry and holding editorial positions in various other reputed journals.

Despite his global recognition, Rao has remained humble and dedicated to the pursuit of knowledge, earning him respect not only as a scientist but also as a person of integrity and character.

Dr. C.N.R. Rao's life story is a testament to the transformative power of science, education, and advocacy. His enduring contributions have not only advanced the frontiers of scientific knowledge but have also shaped the scientific ecosystem in India and inspired countless individuals to pursue excellence in the field of science and research.

3.7 Sir Acharya Prafulla Chandra Ray

Acharya Prafulla Chandra Ray was a renowned Indian chemist, educationist, and industrialist who made significant contributions to the field of chemistry in the early 20th century. He was born on August 2, 1861, in Khulna, Bengal Presidency (now in Bangladesh), and passed away on June 16, 1944. Here is an overview of his life and contributions:

Education and Early Career: Acharya Prafulla Chandra Ray studied at Edinburgh University in Scotland, where he earned a B.Sc. degree in 1881 and a D.Sc. degree in 1887. He was the first Indian to have a Doctorate in Science. He also pursued higher studies in Germany.

Establishment of Bengal Chemical and Pharmaceutical Works: In 1892, Ray founded the Bengal Chemical and Pharmaceutical Works in Kolkata. It was one of the first pharmaceutical companies in India and played a significant role in the growth of the chemical industry in the country.

Chemical Research: Acharya Prafulla Chandra Ray's contributions to chemical research were extensive. He made notable advancements in the synthesis of various organic compounds and was particularly known for his work on mercurous nitrite and the chemistry of oximes.

Foundation of Bengal Chemical Works Laboratory: In 1902, Ray established the

Bengal Chemical Works Laboratory to encourage research in chemistry. The laboratory became a hub for pioneering research and played a crucial role in training young Indian scientists.

Educational Reforms: Apart from his industrial and research contributions, Ray was also involved in educational reforms. He advocated for scientific education and founded the University Science College in Kolkata in 1916, which later became the renowned Rajabazar Science College.

Academic Career: Ray served as the Honorary Professor of Chemistry at the University of Calcutta. His efforts to promote research and education in chemistry were widely recognized.

Scientific Societies: Ray was actively involved in various scientific societies and associations. He played a key role in the formation of the Indian Chemical Society in 1922, serving as its president multiple times.

Recognition and Awards: Acharya Prafulla Chandra Ray received several honors for his contributions to science and industry. He was knighted in 1919 and became the first Indian to be elected as the President of the Indian Science Congress in 1920.

Legacy: Acharya Prafulla Chandra Ray is remembered as one of the pioneers of modern chemical research in India. His contributions to both industry and academia laid the foundation for the growth of the chemical sciences in the country. The Bengal Chemical Works, under his guidance, became a significant institution in the pharmaceutical and chemical sectors.

Literary Contributions: Apart from his scientific pursuits, Ray was also a prolific writer. He authored several books, including "A History of Hindu Chemistry from the Earliest Times to the Middle of the Sixteenth Century."

Acharya Prafulla Chandra Ray's multifaceted contributions to chemistry, education, and industry have left an enduring impact on the scientific landscape of India. His legacy continues to inspire generations of scientists and researchers in the country.

3.8 Dr. Meghnad Saha

Dr. Meghnad Saha was a prominent Indian astrophysicist and mathematician, best known for his groundbreaking work in astrophysics and his formulation of the Saha ionization equation. He made significant contributions to the understanding of stellar spectra and the thermal ionization of elements in stellar atmospheres. Here is an overview of his life and contributions:

Early Life and Education: Meghnad Saha was born on October 6, 1893, in Shaoratoli, a village near Dhaka in present-day Bangladesh. He studied at Dhaka College and later at Presidency College in Kolkata. His academic prowess was evident early on, and he excelled in mathematics.

Cambridge and Berlin: Saha received a scholarship to study at the University of Cambridge in the United Kingdom, where he completed his Tripos in Mathematics in 1913. Subsequently, he pursued a Ph.D. in astrophysics at the University of Berlin under the guidance of renowned physicist Max Planck.

Saha Ionization Equation: Meghnad Saha's most significant contribution to astrophysics is the formulation of the Saha ionization equation, which he published in 1920. The equation describes the ionization state of elements in a gas under thermal equilibrium conditions, particularly in stellar atmospheres. It provided a theoretical framework for understanding the spectral features observed in stars.

Indian Association for the Cultivation of Science (IACS): Upon his return to India, Saha joined the Indian Association for the Cultivation of Science (IACS) in Kolkata. He served as the director of IACS from 1952 to 1956.

Contributions to Astrophysics: Saha's work extended beyond the Saha ionization equation. He made significant contributions to the study of stellar spectra, the structure of stars, and the physical conditions in stellar atmospheres. His research played a crucial role in advancing astrophysical understanding globally.

Thermal Ionization: Saha's research on thermal ionization provided insights into the composition of stars and the conditions prevailing in their atmospheres. This work laid the foundation for the development of astrophysics as a distinct field of study.

Recognition and Honors: Meghnad Saha received numerous accolades for his scientific contributions. He was elected a Fellow of the Royal Society in 1927, and he received the Padma Bhushan, one of India's highest civilian awards, in 1954.

Political Involvement: Saha was actively involved in the socio-political landscape of India. He was a member of the Indian National Congress and served as a Member of Parliament. He also played a key role in scientific planning and policymaking in independent India.

Legacy: Meghnad Saha's legacy extends beyond his scientific achievements. He played a crucial role in the development of scientific institutions in India and contributed to the establishment of the Council of Scientific and Industrial Research (CSIR). The Saha Institute of Nuclear Physics in Kolkata was named in his honor.

Demise: Meghnad Saha passed away on February 16, 1956, leaving behind a lasting legacy in the fields of astrophysics and scientific research in India.

Dr. Meghnad Saha's pioneering work laid the foundation for the study of astrophysics in India and contributed significantly to our understanding of the physical processes occurring in the vast reaches of the universe. His scientific and political contributions have left an indelible mark on the scientific community and the nation.

3.9 Dr. PC Mahalanobis

Dr. Prasanta Chandra Mahalanobis, often referred to as P.C. Mahalanobis, was a renowned Indian statistician and applied mathematician who played a crucial role in the development of statistical science in India. Here is an overview of his life and contributions:

Early Life and Education: Prasanta Chandra Mahalanobis was born on June 29, 1893, in Calcutta (now Kolkata), India. He came from a distinguished family; his father, Prabodh Chandra, was a distinguished academician and the first Indian to be appointed as a fellow of the Royal Society.

Education in England: Mahalanobis pursued his education in England, studying at King's College, Cambridge. He graduated with honors in Physics and later studied at the University of London.

Indian Statistical Institute (ISI): In 1931, Mahalanobis founded the Indian Statistical Institute (ISI) in Kolkata, which became a pioneering institution for statistical research and training. ISI played a crucial role in advancing statistical methods and applications in various fields.

Mahalanobis Distance: One of his significant contributions to statistics is the development of the Mahalanobis Distance, a measure used to quantify the distance between a point and a distribution. This method has applications in various fields, including pattern recognition and clustering.

National Sample Survey (NSS): Mahalanobis played a pivotal role in the establishment of the National Sample Survey (NSS) in 1950. The NSS has been instrumental in collecting data on socio-economic indicators, contributing to evidence-based policy formulation in India.

Planning Commission and Economic Planning: Mahalanobis was appointed as the Honorary Statistical Advisor to the Government of India and was a key figure in the Planning Commission. He played a crucial role in the formulation of India's Second Five-Year Plan (1956–1961), emphasizing the importance of industrialization and economic planning.

Economic Planning Model: Mahalanobis advocated for a model of economic development that focused on the growth of heavy industries, which became known as the "Mahalanobis Model." His approach aimed at reducing dependence on foreign imports and promoting self-sufficiency.

Contributions to Anthropometry: Mahalanobis made contributions to anthropometry, the scientific study of measurements and proportions of the human body. His work in this area had applications in designing ergonomic systems and addressing health and nutrition concerns.

Recognition and Honors: P.C. Mahalanobis received numerous honors for his contributions, including being elected as a Fellow of the Royal Society in 1945. He was also the president of the Indian Science Congress in 1947.

Demise: Dr. Prasanta Chandra Mahalanobis passed away on June 28, 1972, leaving behind a legacy of statistical research, economic planning, and institutional development.

P.C. Mahalanobis is remembered not only as a statistician but also as a visionary planner and institution-builder. His contributions have had a lasting impact on the development of statistical methods, economic planning, and the application of statistics to various domains in India. The Indian Statistical Institute and the National Sample Survey are enduring monuments to his pioneering work.

3.10 Srinivasa Ramanujan

Srinivasa Ramanujan was a self-taught Indian mathematician who made substantial contributions to mathematical analysis, number theory, infinite series, and continued fractions. His work is characterized by its deep insights, originality, and a unique blend of intuition and formal proof. Here is an overview of his life and contributions:

Early Life: Srinivasa Ramanujan was born on December 22, 1887, in Erode, Tamil Nadu, India. From a young age, he showed an extraordinary aptitude for mathematics, independently discovering theorems and developing his own mathematical methods.

Educational Background: Despite facing financial hardships, Ramanujan gained admission to the Government Arts College in Kumbakonam. However, due to his focus on self-study and unorthodox methods, he struggled with formal education.

Mathematical Achievements: Ramanujan independently discovered a multitude of results in number theory, modular forms, elliptic functions, and infinite series. His work included novel solutions to mathematical problems that had puzzled mathematicians for centuries.

Correspondence with G.H. Hardy: In 1913, Ramanujan wrote to the eminent British mathematician G.H. Hardy, enclosing a collection of his theorems. Recognizing the brilliance of Ramanujan's work, Hardy invited him to Cambridge, where Ramanujan eventually spent five years collaborating with Hardy and others.

Ramanujan-Hardy Number: The famous number 1729 is known as the "Ramanujan-Hardy number." It is the smallest positive integer expressible as the sum of two cubes in two different ways: $1729=1^3+12^3=9^3+10^3$. This number became well-known due to a story involving a visit by Hardy to Ramanujan in the hospital.

Contributions to Number Theory: Ramanujan made groundbreaking contributions to number theory, including his work on highly composite numbers, partition functions, and the distribution of prime numbers. His formula for the partition function remains influential.

Modular Forms and Elliptic Functions: Ramanujan's discoveries in the theory of modular forms and elliptic functions significantly influenced the field of mathematics. His findings laid the groundwork for further developments in these areas.

Illness and Return to India: Ramanujan faced health issues during his time in England, possibly exacerbated by the cold climate. In 1919, he returned to India, where he continued to work on mathematics. Unfortunately, his health deteriorated rapidly.

Legacy and Influence: Ramanujan's work has had a profound impact on various branches of mathematics. Mathematicians around the world continue to explore and build upon his discoveries. The "Ramanujan conjecture" and the "Ramanujan-Hardy-Littlewood circle method" are enduring aspects of his legacy.

Death: Srinivasa Ramanujan passed away on April 26, 1920, at the young age of 32. The exact cause of his death remains uncertain but is often attributed to tuberculosis.

Ramanujan's life and work exemplify the power of raw mathematical intuition and the potential for extraordinary contributions to the field, even in the absence of formal education. His story continues to inspire mathematicians and serves as a testament to the universality and beauty of mathematical discovery.

3.11 Dr. Har Gobind Khorana

Dr. Har Gobind Khorana was an Indian-American biochemist renowned for his groundbreaking contributions to the field of genetics, particularly in deciphering the genetic code and understanding how information in DNA is translated into proteins. His work laid the foundation for advancements in molecular biology and biotechnology. Here is an overview of his life and achievements:

Early Life and Education: Har Gobind Khorana was born on January 9, 1922, in Raipur, British India (now in Pakistan). Despite facing financial challenges, he pursued his education, earning his Bachelor's degree in Lahore and later obtaining a Ph.D. in Organic Chemistry from the University of Liverpool in 1948.

Postdoctoral Research: After completing his Ph.D., Khorana moved to Switzerland for postdoctoral research. Subsequently, he worked at institutions in England and Canada, where he began his pioneering work in the field of molecular biology.

Nirenberg and Khorana: The Genetic Code: In the 1960s, Har Gobind Khorana collaborated with Marshall W. Nirenberg to crack the genetic code, revealing how the sequence of nucleotide bases in DNA instructs the synthesis of proteins. This groundbreaking work earned them the Nobel Prize in Physiology or Medicine in 1968.

RNA and Protein Synthesis: Khorana's research expanded our understanding of the role of RNA in protein synthesis. He synthesized the first artificial gene and demonstrated the feasibility of constructing genes in the laboratory.

Nobel Prize in Physiology or Medicine (1968): Khorana shared the Nobel Prize with Nirenberg and Robert W. Holley for their contributions to deciphering the genetic code and establishing the link between the nucleotide sequence of DNA and the amino acid sequence of proteins.

DNA Synthesis and Nobel Lecture: In his Nobel Lecture, Khorana discussed his work on the synthesis of DNA and its implications for understanding the genetic code. He emphasized the importance of molecular biology in advancing medical research.

Academic Positions and Institutions: Throughout his career, Khorana held various academic positions at institutions such as the University of Wisconsin-Madison and the Massachusetts Institute of Technology (MIT). He made significant contributions to the field of biotechnology.

Artificial Gene Synthesis: Khorana continued to make important contributions to DNA synthesis. In 1972, he and his team successfully synthesized the first artificial gene, paving the way for advancements in genetic engineering.

Honors and Awards: In addition to the Nobel Prize, Khorana received numerous awards and honors for his scientific contributions, including the National Medal of Science and the Padma Vibhushan, one of India's highest civilian awards.

Later Life: Har Gobind Khorana continued his research until his retirement. He passed away on November 9, 2011, in Concord, Massachusetts, leaving behind a legacy of groundbreaking contributions to molecular biology and genetics.

Dr. Har Gobind Khorana's work laid the foundation for the understanding of genetic information and paved the way for subsequent advancements in genetic engineering and biotechnology. His legacy continues to inspire scientists and researchers in the field of molecular biology.

3.12 Dr. Subrahmanyan Chandrashekhar

Dr. Subrahmanyan Chandrasekhar, often referred to as Chandra, was an Indian-American astrophysicist known for his significant contributions to the understanding of stellar structure, black holes, and the evolution of stars. His work earned him numerous accolades, including the Nobel Prize in Physics. Here is an overview of his life and achievements:

Early Life and Education: Subrahmanyan Chandrasekhar was born on October 19, 1910, in Lahore, British India (now in

Pakistan). He came from a family of scholars and displayed exceptional mathematical abilities from a young age. He studied at Presidency College in Chennai (Madras) and later pursued postgraduate studies at the University of Cambridge in England.

Chandrasekhar Limit: In 1930, at the age of 19, Chandra formulated the Chandrasekhar limit, which describes the maximum mass of a stable white dwarf star. This groundbreaking work became fundamental to the understanding of stellar evolution.

Controversy and Eddington: When Chandrasekhar presented his limit at a Royal Astronomical Society meeting, Sir Arthur Eddington, a prominent astrophysicist of the time, resisted the concept. However, Chandra's calculations were eventually recognized, and the Chandrasekhar limit became a cornerstone in astrophysics.

Career in the United States: Chandrasekhar moved to the United States in 1937, where he began teaching and conducting research. He accepted a position at the University of Chicago, where he spent the majority of his academic career.

Contributions to Astrophysics: Chandrasekhar made significant contributions to various areas of astrophysics, including the theory of white dwarfs, the dynamics of stellar systems, and the mathematical theory of black holes.

Nobel Prize in Physics (1983): In 1983, Chandrasekhar was awarded the Nobel Prize in Physics for his work on the physical processes important to the structure and evolution of stars. The Nobel Committee cited his discoveries related to the mathematical theory of black holes.

Publications and Books: Chandrasekhar authored numerous scientific papers and books throughout his career. His book "Principles of Stellar Dynamics" and "Introduction to the Study of Stellar Structure" are considered seminal works in astrophysics.

Accretion Theory and Radiative Transfer: Chandrasekhar's research also extended to accretion theory, which explains the accumulation of mass by celestial objects, and radiative transfer, the study of the interaction of radiation with matter.

Later Life and Legacy: Chandrasekhar continued to be active in research and teaching until his retirement. He passed away on August 21, 1995, in Chicago. His legacy includes numerous honors, awards, and the lasting impact of his contributions on the field of astrophysics.

Chandrasekhar's Limit Revisited: In the later part of the 20th century and into the 21st century, observations and theoretical developments confirmed the existence of objects that surpassed Chandrasekhar's limit, leading to further discussions and advancements in our understanding of stellar evolution.

Dr. Subrahmanyan Chandrasekhar's work laid the groundwork for modern astrophysics and significantly advanced our understanding of the structure and behavior of celestial bodies. His contributions continue to influence research in the field, and his legacy as a brilliant astrophysicist endures.

3.13 SS Abhyankar

Shreeram Shankar Abhyankar, known as S.S. Abhyankar, was an Indian-American mathematician known for his contributions to algebraic geometry and his work on algebraic varieties, singularities, and resolution of singularities. He made significant advancements in the understanding of algebraic equations and their geometric representations. Here is an overview of his life and work:

Early Life and Education: S.S. Abhyankar was born on July 22, 1930, in Ujjain, British India. He completed his undergraduate studies at the University of Mumbai (then Bombay) and later earned a Ph.D. in mathematics from Harvard University in 1955 under the guidance of Oscar Zariski.

Contributions to Algebraic Geometry: Abhyankar's early work focused on algebraic geometry, particularly on topics related to algebraic varieties, singularities, and resolution of singularities. His work had a profound impact on the field, and he made significant contributions to the study of plane algebraic curves.

Abhyankar's Conjecture: Abhyankar formulated the famous "Abhyankar's Conjecture," which is related to the theory of finite groups acting on algebraic varieties. The conjecture became an influential topic in algebraic geometry.

Resolution of Singularities: Abhyankar made important contributions to the resolution of singularities, a problem in algebraic geometry that deals with finding a "nicer" space that is homeomorphic to a given space but has simpler geometric properties.

Mathematics Journals and Editorial Work: Abhyankar served as an editor for several prestigious mathematical journals. He contributed to the development and dissemination of mathematical research through his editorial roles.

Academic Positions: Abhyankar held various academic positions throughout his career. He taught at institutions such as Cornell University and Purdue University, making significant contributions to the mathematical community through his research and mentorship of students.

Awards and Recognition: S.S. Abhyankar received several awards and honors for his contributions to mathematics. He was elected as a fellow of the American Academy of Arts and Sciences and the National Academy of Sciences.

Later Life: In the later part of his career, Abhyankar continued his research and academic pursuits. He remained active in the mathematical community, attending conferences, and collaborating with other researchers.

Demise: S.S. Abhyankar passed away on November 2, 2012, in West Lafayette, Indiana, USA. His contributions to algebraic geometry and singularities remain influential, and his work continues to be studied and built upon by mathematicians worldwide.

S.S. Abhyankar's work significantly impacted algebraic geometry, and his contributions to the understanding of algebraic varieties and singularities have left a lasting legacy in the field of mathematics. His conjectures and results continue to inspire further research and advancements in algebraic geometry.

3.14 Dr. Birbal Sahni

Dr. Birbal Sahni was an Indian paleobotanist who made significant contributions to the field of paleobotany and plant morphology. He was also a pioneer in the establishment of scientific institutions in India. Here is an overview of his life and contributions:

Early Life and Education: Birbal Sahni was born on November 14, 1891, in Bhera, British India (now in Pakistan). He studied at Government College in Lahore and later pursued his D.Sc. at Emmanuel College, Cambridge, under the guidance of Professor A.C. Seward.

Contribution to Paleobotany: Dr. Birbal Sahni made substantial contributions to paleobotany, the study of ancient plants based on fossils. His research focused on the Mesozoic and Cenozoic plant fossils found in India. Sahni's work provided insights into the evolutionary history of plant life.

Establishment of the Birbal Sahni Institute of Palaeobotany: In 1946, Birbal Sahni founded the Birbal Sahni Institute of Palaeobotany in Lucknow, India. The institute became a leading center for research in paleobotany, contributing to the understanding of plant evolution over geological time.

Research in Plant Morphology: Sahni's research extended to plant morphology, and he made significant contributions to understanding the structure and form of plants. His studies on the anatomy of fossil and living plants added valuable knowledge to the field.

Scientific Institutions: Apart from the Birbal Sahni Institute of Palaeobotany, Sahni played a crucial role in the establishment of other scientific institutions in India. He was associated with the National Institute of Sciences of India (now Indian National Science Academy) and the Indian Botanical Society.

Academic Leadership: Birbal Sahni held various academic positions during his career. He served as the Director of the Botanical Survey of India and was a Professor of Botany at the University of Lucknow.

Honors and Awards: Dr. Sahni received numerous honors and awards for his contributions to science. He was elected as a Fellow of the Royal Society (London) and received the prestigious Copley Medal in 1946.

Educational Reforms: Birbal Sahni was involved in educational reforms and emphasized the importance of scientific research and education in India. His efforts contributed to the growth of scientific institutions and research culture in the country.

Demise: Birbal Sahni passed away on April 10, 1949, at the age of 57. Despite his relatively short life, his contributions to paleobotany and scientific institutions in India had a lasting impact.

Legacy: The Birbal Sahni Institute of Palaeobotany continues to be a leading center for paleobotanical research in India. Dr. Birbal Sahni's legacy lives on through his contributions to the understanding of plant evolution and the institutions he established, fostering scientific research in the country.

Dr. Birbal Sahni's work in paleobotany and plant morphology has left an enduring impact on the field of plant sciences. His dedication to scientific research and the establishment of institutions has contributed significantly to the growth of botanical studies in India.

3.15 Legacy and Inspiration

As we conclude this chapter, the legacy of these Indian science giants reverberates through the corridors of time. Their contributions have not only advanced scientific knowledge but have also served as a wellspring of inspiration for future generations. The subsequent chapters will build upon this legacy, exploring the intersection of traditional knowledge and modern innovations in the dynamic landscape of Indigenous Technologies.

As we reflect on the legacies of Indian science giants like Sir Jagadish Chandra Bose and Dr. Homi Bhabha, their impact echoes through the annals of time. These pioneers have left an indelible mark on the scientific community, shaping not only the understanding of fundamental principles but also influencing the trajectory of technological advancements. Their stories serve as beacons of inspiration for future generations of scientists and innovators.

Moving forward, the narrative will delve into the intersection of traditional knowledge and modern innovations in the dynamic landscape of Indigenous Technologies. This exploration is particularly pertinent as it illuminates the symbiotic relationship between age-old wisdom and contemporary scientific practices. Indigenous Technologies encompass a diverse array of innovations rooted in local knowledge, cultural practices, and a deep understanding of the environment.

The subsequent chapters will unravel stories of inventors and innovators who, drawing upon the rich tapestry of traditional wisdom, have developed solutions that bridge the gap between the past and the future. These stories showcase how indigenous communities have been at the forefront of sustainable practices, environmental conservation, and resource utilization. From agricultural techniques passed down through generations to eco-friendly technologies grounded in traditional craftsmanship, the exploration of Indigenous Technologies promises a journey into the heart of innovation deeply rooted in cultural heritage.

Moreover, the intersection of traditional knowledge and modern science represents a holistic approach to problem-solving. By integrating the best of both worlds, societies can address contemporary challenges while preserving cultural identities and ecological balance. The upcoming chapters will unfold narratives that exemplify how Indigenous Technologies contribute not only to scientific and technological progress but also to the sustainable development of communities.

In the pages to come, the exploration of Indigenous Technologies will be a celebration of ingenuity, resilience, and the enduring spirit of innovation. It will showcase how the lessons of the past continue to guide and inspire the creation of a more sustainable and harmonious future, bridging the temporal gap between the wisdom of yesteryears and the possibilities of tomorrow.

CHAPTER FOUR

Indigenous Technologies in Agriculture

4.1 The Crucial Role of Agriculture in India's Development

Agriculture has been the backbone of India's economy, providing livelihoods to a significant portion of the population. This chapter delves into the application of Indigenous Technologies within the agricultural sector, emphasizing their pivotal role in shaping the nation's development trajectory.

Certainly, the role of agriculture in India's economy has been significant, and the utilization of Indigenous Technologies in the agricultural sector has played a crucial role in shaping the country's development trajectory.

1. Traditional Agricultural Practices:

India has a rich history of traditional and indigenous agricultural practices that have been passed down through generations. These practices are often well-suited to local climates and conditions.

Techniques such as crop rotation, mixed cropping, and organic farming have been integral to Indian agriculture for centuries, contributing to sustainable and environmentally friendly practices.

2. Water Management:

Indigenous technologies for water conservation and management, such as traditional water harvesting systems like 'Johads' and 'Kunds,' have been employed to efficiently store rainwater in arid regions.

These methods are cost-effective and have proven to be crucial for mitigating water scarcity issues in many parts of the country.

3. Seed Preservation and Breeding:

Farmers have traditionally engaged in seed preservation practices, selecting and saving seeds from the best-performing crops for the next planting season.

Indigenous breeding techniques have led to the development of region-specific crop varieties, adapted to local conditions and providing better yields.

4. Livestock Management:

Indigenous knowledge in livestock management includes practices like selective breeding, healthcare through traditional medicine, and sustainable grazing patterns.

This knowledge has been essential for the well-being of livestock, which in turn supports the livelihoods of many rural communities.

5. Pest Management:

Traditional pest management practices involve the use of natural pesticides derived from locally available plants.

Crop diversification and intercropping, as per indigenous wisdom, help in reducing vulnerability to pests and diseases.

6. Community-based Farming:

Many indigenous practices emphasize community-based farming, where farmers work together, share resources, and collectively make decisions for the benefit of the entire community.

This communal approach fosters a sense of unity and support among farmers, contributing to sustainable agricultural development.

7. Modern Integration:

The integration of modern technologies with indigenous knowledge has led to innovations like precision farming, where data and technology are used to optimize crop yields without compromising on traditional practices.

In conclusion, the application of Indigenous Technologies in agriculture has been fundamental to India's economic backbone. Recognizing the value of traditional knowledge alongside modern advancements is crucial for achieving sustainable and inclusive agricultural development in the country. It not only preserves cultural heritage but also ensures that agriculture remains a resilient and productive sector, providing livelihoods for a significant portion of the population.

4.2 Traditional Farming Practices: An Ancient Wisdom

The narrative begins by unraveling the tapestry of traditional farming practices deeply rooted in India's agricultural heritage. From organic farming techniques to crop rotation and water harvesting methods, this section explores the timeless wisdom embedded in traditional agriculture and its relevance in the contemporary context.

As we embark on this journey through India's agricultural heritage, we find ourselves peeling back the layers of time to reveal the rich tapestry of traditional farming practices. Rooted deep in the soil of the subcontinent, these age-old techniques form the backbone of an agricultural legacy that has sustained communities for generations.

Organic farming, a practice as old as agriculture itself, takes center stage in this exploration. Passed down through the ages, the wisdom of cultivating the land without synthetic chemicals has been a hallmark of Indian farming. Traditional farmers, attuned to the rhythms of nature, understood the delicate balance that exists within ecosystems. They worked in harmony with the soil, cultivating crops without disturbing the delicate equilibrium that sustains life.

Crop rotation, another ancient technique woven into the fabric of Indian agriculture, emerges as a key thread in this narrative. Farmers, guided by an intuitive understanding of the land, rotated crops to maintain soil fertility and prevent the depletion of essential nutrients. This cyclical dance with nature not only ensured bountiful harvests but also safeguarded the health of the land for future generations.

Water harvesting methods, deeply ingrained in the agricultural practices of yore, find renewed relevance in today's context of climate change and water scarcity. Traditional wisdom recognized the preciousness of water and devised ingenious methods to capture and store rainwater. From ancient stepwells to intricate canal systems, these age-old techniques offer valuable insights for modern-day sustainable water management.

As we unravel the layers of tradition, it becomes evident that the roots of sustainable agriculture run deep in India's farming heritage. The challenges of the contemporary world – climate change, soil degradation, and water scarcity – beckon us to revisit and revitalize these time-tested practices. In doing so, we not only honor the wisdom of our ancestors but also forge a path towards a more sustainable and resilient future for agriculture in India and beyond.

4.3 Precision Farming: Modern Innovations for Enhanced Productivity

As we progress, the chapter highlights the infusion of modern technologies into agriculture, particularly precision farming. The integration of data-driven approaches, IoT (Internet of Things), and remote sensing technologies are explored for their role in optimizing resource usage, enhancing crop yields, and ensuring sustainable agricultural practices.

The agricultural landscape is undergoing a transformation propelled by cutting-edge technologies, and at the forefront of this revolution is precision farming. This modern approach leverages a suite of innovations to optimize every aspect of the farming process, from planting to harvesting, with the goal of enhancing productivity and sustainability.

1. GPS Technology: Precision farming relies heavily on Global Positioning System (GPS) technology. Farmers use GPS to precisely map their fields, enabling accurate navigation of machinery. This technology allows for targeted application of resources, minimizing waste and maximizing efficiency. It also aids in the creation of detailed field maps for better decision-making.

2. Sensors and IoT Devices: Sensors and Internet of Things (IoT) devices play a pivotal role in collecting real-time data from the field. These devices can measure soil moisture levels, nutrient content, and even monitor the health of crops. This wealth of information allows farmers to make data-driven decisions, adjusting inputs such as water, fertilizers, and pesticides precisely where and when they are needed.

3. Drones and Satellite Imaging: Unmanned aerial vehicles (UAVs) or drones equipped with advanced imaging technology provide farmers with high-resolution aerial views of their fields. Satellite imaging complements this by offering broader coverage. These tools assist in identifying crop health, pest infestations, and areas requiring specific attention. Farmers can respond promptly to issues, preventing widespread crop damage.

4. Automated Machinery and Robotics: Precision farming incorporates automated machinery and robotics for various tasks, such as planting, weeding, and harvesting. These machines are equipped with sensors and actuators that enable precise and consistent operations. Autonomous tractors and robotic harvesters reduce labor demands and enhance efficiency.

5. Variable Rate Technology (VRT): VRT enables farmers to apply inputs at variable rates across a field based on specific conditions. This can include adjusting

fertilizer application, seeding density, or irrigation levels. By tailoring inputs to the unique needs of different areas within a field, farmers optimize resource utilization and minimize environmental impact.

6. Data Analytics and Decision Support Systems: The vast amount of data generated by precision farming technologies is processed and analyzed using advanced analytics. Decision support systems provide actionable insights, helping farmers make informed choices about planting strategies, crop rotations, and resource management.

7. Smart Farming Apps: Mobile applications offer farmers user-friendly interfaces to monitor and control various aspects of their operations. These apps can provide real-time weather updates, allow remote control of machinery, and offer insights derived from data analytics.

Precision farming represents a paradigm shift in agriculture, blending age-old practices with cutting-edge technologies. By harnessing these innovations, farmers can not only boost productivity but also contribute to sustainable and resource-efficient agricultural practices for the future.

4.4 Bio-fertilizers and Sustainable Agriculture

The narrative unfolds into the realm of sustainable agriculture, where indigenous technologies play a crucial role in the development and application of bio-fertilizers. This section explores the eco-friendly alternatives to chemical fertilizers, showcasing the benefits of microbial-based fertilizers in promoting soil health, improving nutrient absorption, and reducing environmental impact.

1. Introduction:

Bio-fertilizers are substances containing living microorganisms that enhance soil fertility and plant nutrition. They play a vital role in sustainable agriculture by promoting environmentally friendly farming practices.

2. Types of Bio-fertilizers:

Nitrogen-Fixing Bacteria: Rhizobium and Azotobacter fix atmospheric nitrogen into a form usable by plants, reducing the need for synthetic nitrogen fertilizers.

Phosphate-Solubilizing Bacteria: Microorganisms like mycorrhizal fungi and phosphate-solubilizing bacteria enhance phosphorus availability to plants.

Potash Mobilizing Bacteria: Bacteria like Bacillus mucilaginosus help in the release of potassium from minerals in the soil.

3. Benefits of Bio-fertilizers:

Environmentally Friendly: Bio-fertilizers reduce the reliance on chemical fertilizers, minimizing environmental pollution and soil degradation.

Improved Soil Structure: They contribute to soil health by enhancing its structure, water-holding capacity, and nutrient content.

Symbiotic Relationships: Nitrogen-fixing bacteria form symbiotic relationships with leguminous plants, fostering a mutually beneficial exchange of nutrients.

Cost-Effective: Bio-fertilizers can be a cost-effective alternative to synthetic fertilizers, particularly in the long term.

4. Contribution to Sustainable Agriculture:

Reduced Chemical Dependency: The use of bio-fertilizers reduces the dependence on chemical inputs, promoting sustainable and organic farming practices.

Biodiversity and Soil Health: Bio-fertilizers contribute to the overall health of the soil, supporting diverse microbial populations and preventing soil degradation.

Water Conservation: Improved soil structure and water-holding capacity resulting from bio-fertilizer use contribute to water conservation in agriculture.

5. Challenges and Considerations:

Specificity: Some bio-fertilizers are specific to certain crops or conditions, requiring careful selection and application.

Storage and Shelf Life: Maintaining the viability of living microorganisms in bio-fertilizers during storage can be challenging.

Integration with Other Practices: Successful implementation often requires integration with other sustainable agricultural practices like crop rotation and cover cropping.

6. Future Prospects:

Research and Innovation: Ongoing research is essential for developing new strains of bio-fertilizers and optimizing their application for various crops and agroecological conditions.

Awareness and Adoption: Promoting awareness among farmers about the benefits of bio-fertilizers and providing support for their adoption is crucial for widespread use.

7. Government Initiatives:

Governments can play a role in incentivizing the use of bio-fertilizers through subsidies, education programs, and research funding.

In conclusion, the use of bio-fertilizers is a key component of sustainable agriculture, contributing to soil health, reducing environmental impact, and supporting long-term food security. Continued research, education, and supportive policies are essential for maximizing the benefits of bio-fertilizers in the context of sustainable agricultural practices.

4.5 Water Management: Harnessing Indigenous Wisdom

Water scarcity poses a significant challenge to agriculture. The chapter delves into indigenous water management techniques that have sustained farming communities for centuries. From traditional rainwater harvesting methods to community-driven irrigation systems, this section explores how indigenous technologies contribute to efficient water usage in agriculture.

1. Traditional Water Harvesting Systems:

Indigenous communities often developed ingenious water harvesting systems suited to local conditions. Examples include 'Johads' in Rajasthan, 'Kulhs' in Himachal Pradesh, and 'Eri' in Tamil Nadu.

These systems are designed to capture and store rainwater, preventing runoff and ensuring a sustainable water supply for agriculture and community needs.

2. Traditional Irrigation Techniques:

Indigenous communities have mastered efficient irrigation methods, such as 'Ahar-Pyne' systems in eastern India, which involve diverting water through a network of channels to irrigate fields.

Traditional knowledge often emphasizes the importance of timing and water distribution, optimizing agricultural productivity.

3. Sacred Groves and Watershed Management:

Many indigenous cultures consider certain areas as sacred groves, protecting water sources and maintaining biodiversity.

Indigenous wisdom includes practices like watershed management, ensuring the preservation of forests and natural water catchment areas.

4. Community-led Water Conservation:

Indigenous communities often adopt a communal approach to water management, involving collective decision-making and shared responsibility.

Practices such as 'Eco-village' models focus on community-led water conservation, emphasizing the sustainable use of water resources.

5. Intercropping and Crop Rotation:

Traditional farming practices often involve intercropping and crop rotation, which not only enhance soil fertility but also contribute to efficient water use.

These methods reduce the risk of water-intensive monoculture and promote resilience in the face of variable precipitation.

6. Seasonal Water Storage:

Indigenous communities are adept at seasonal water storage, using traditional structures like 'Talabs' and 'Bandharas' to store water during periods of abundance for use during dry seasons.

This ensures a continuous water supply for agriculture and sustains ecosystems.

7. Indigenous Knowledge in Drought Management:

Traditional knowledge systems include strategies for coping with drought, such as selecting drought-resistant crops, practicing rainwater harvesting, and implementing water-use restrictions during dry periods.

8. Modern Integration of Indigenous Wisdom:

There is a growing recognition of the value of integrating modern technologies with indigenous water management practices.

Remote sensing, data analytics, and community-based monitoring can enhance the effectiveness of traditional water management approaches.

9. Knowledge Transfer and Preservation:

Efforts should be made to document and preserve indigenous water management practices, facilitating knowledge transfer to future generations.

Collaborative initiatives between indigenous communities and scientific institutions can help bridge traditional wisdom with contemporary research.

10. Policy Support:

Governments can play a crucial role by acknowledging and incorporating indigenous water management practices into policies and programs.

Supportive policies can incentivize the adoption of traditional water conservation methods and ensure the sustainability of water resources.

In conclusion, harnessing indigenous wisdom in water management is essential for building resilient and sustainable water systems. Integrating traditional knowledge with modern innovations can lead to effective and holistic approaches to address current and future water challenges, ensuring the well-being of communities and the environment.

4.6 Agro-Technology Startups: Nurturing Innovation

The exploration extends to the vibrant ecosystem of agro-technology startups that are leveraging indigenous knowledge and cutting-edge technologies. This section showcases examples of startups developing smart farming solutions, AI-driven pest control, and other innovations aimed at revolutionizing the agricultural landscape in India.

The agricultural sector is experiencing a wave of innovation driven by agro-technology startups. These dynamic enterprises are bringing forth novel solutions to address challenges in farming, resource management, and sustainability. Nurturing innovation, agro-tech startups are reshaping the landscape of agriculture and contributing to the development of a more resilient and efficient food production system.

1. Precision Agriculture Solutions: Many startups focus on precision agriculture technologies, offering farmers tools for data-driven decision-making. These solutions incorporate sensors, drones, and satellite imaging to gather information about soil health, crop conditions, and environmental factors. By providing farmers with actionable insights, these startups empower them to optimize resource use, reduce waste, and enhance overall productivity.

2. Sustainable Farming Practices: Startups are championing sustainable farming practices by

developing technologies that promote eco-friendly approaches. This includes the use of organic inputs, crop rotation planning, and integrated pest management systems. These innovations not only contribute to environmental conservation but also support the growing demand for sustainably produced food.

3. Agri-FinTech Platforms: Agri-FinTech startups are leveraging technology to address financial challenges in agriculture. These platforms provide farmers with access to credit, insurance, and other financial services. By utilizing data analytics and machine learning, these startups assess risks and offer tailored financial solutions, promoting financial inclusion in rural areas.

4. Supply Chain Optimization: Startups are innovating in supply chain management to reduce post-harvest losses and improve overall efficiency. Technologies such as blockchain, IoT sensors, and data analytics are employed to create transparent and

traceable supply chains. This ensures that agricultural products reach consumers with minimal waste and maximum freshness.

5. Aquaculture and Vertical Farming: Innovations in aquaculture and vertical farming are gaining traction. Startups are developing sustainable aquaculture practices and urban farming solutions that maximize space and reduce the environmental footprint of food production. These technologies aim to address the challenges of land scarcity and water use efficiency.

6. Farm Robotics and Automation: Startups are introducing robotics and automation into farming operations to address labor shortages and increase efficiency. Autonomous tractors, robotic harvesters, and AI-driven sorting systems are examples of technologies that reduce the reliance on manual labor while improving the precision and speed of various tasks.

7. Agro-Educational Platforms: Some startups focus on agro-educational platforms that

provide farmers with training and information. These platforms utilize multimedia, mobile apps, and online courses to disseminate knowledge about modern agricultural practices, technology adoption, and market trends.

8. Weather Forecasting and Climate Resilience: Startups are developing advanced weather forecasting tools and climate-resilient solutions to help farmers mitigate the impact of changing weather patterns. This includes early warning systems for extreme weather events, as well as technologies that adapt farming practices to evolving climate conditions.

As these agro-tech startups continue to flourish, they not only drive innovation within the agriculture sector but also contribute to the broader goal of creating a more sustainable, efficient, and resilient food production ecosystem. Nurturing these startups becomes crucial for fostering continuous advancements in agriculture and addressing the evolving challenges faced by the global farming community.

4.7 Challenges and Opportunities in Agricultural Innovation

While indigenous technologies bring forth a myriad of opportunities, the chapter does not shy away from addressing the challenges faced in their adoption. Issues such as access to technology, farmer education, and policy support are discussed, offering insights into potential solutions and pathways for overcoming these obstacles.

Challenges:

Resource Constraints:

Financial Limitations: Many small-scale farmers may lack the financial resources to adopt new technologies or invest in innovative agricultural practices.

Limited Infrastructure: In some regions, inadequate infrastructure, such as lack of irrigation or storage facilities, hampers the adoption of innovative practices.

Technological Accessibility:

Digital Divide: Unequal access to technology, especially in rural areas, poses a challenge. Farmers may struggle to access and utilize digital tools and precision agriculture technologies.

Knowledge and Training:

Awareness: Limited awareness and knowledge about innovative practices hinder their adoption. Providing education and training to farmers is crucial for successful implementation.

Capacity Building: Building the capacity of farmers to understand and implement new technologies is essential.

Policy and Regulatory Issues:

Lack of Supportive Policies: Inconsistent or inadequate policies can impede the adoption of innovative agricultural practices. There may be a need for policies that incentivize and support innovation.

Regulatory Barriers: Cumbersome regulations may slow down the approval and adoption of new technologies, such as genetically modified crops.

Climate Change and Variability:

Unpredictable Weather Patterns: Climate change introduces uncertainties and challenges in planning agricultural activities. Innovations need to be adaptable to changing climatic conditions.

Extreme Events: The increasing frequency of extreme weather events poses a threat to agricultural productivity, requiring resilient and innovative solutions.

Opportunities:

Technological Advancements:

Precision Agriculture: Technologies such as GPS-guided tractors, drones, and sensors enable precision agriculture, optimizing resource use and enhancing productivity.

Biotechnology: Advances in biotechnology offer opportunities for crop improvement, disease resistance, and increased nutritional content.

Data-driven Agriculture:

Big Data and Analytics: Harnessing big data can provide valuable insights into crop management, pest control, and market trends, improving decision-making for farmers.

Internet of Things (IoT): IoT devices can be used for real-time monitoring of crop conditions, soil health, and irrigation needs.

Sustainable Practices:

Agroecology: Integrating ecological principles into agriculture promotes sustainable and environmentally friendly practices.

Organic Farming: Growing consumer demand for organic products creates opportunities for farmers to adopt organic farming practices.

Market Linkages:

E-commerce Platforms: Online platforms connect farmers directly to consumers, improving market access and reducing dependency on traditional supply chains.

Value Addition: Innovations in food processing and value addition can enhance the income of farmers.

Collaboration and Partnerships:

Public-Private Partnerships: Collaborations between government, private sector, and research institutions can facilitate the development and adoption of innovative agricultural technologies.

Knowledge Exchange: Platforms for sharing best practices and knowledge transfer can accelerate the adoption of innovations across regions.

Climate-smart Agriculture:

Drought-resistant Crops: Research and development of crops that are resistant to drought and other climatic stresses offer opportunities for sustainable agriculture.

Water Management Technologies: Innovations in water-efficient irrigation systems and rainwater harvesting can help address water scarcity challenges.

In conclusion, addressing the challenges and capitalizing on the opportunities in agricultural innovation require a holistic approach involving stakeholders at various levels. This includes the development of supportive policies, investment in research and development, and the promotion of knowledge dissemination and capacity building among farmers. By overcoming these challenges and embracing opportunities, agriculture can become more resilient, sustainable, and capable of meeting the growing demands for food in the future.

4.8 Future Outlook: A Sustainable Agricultural Landscape

As we conclude this chapter, the focus turns towards the future of agriculture in India, where the integration of indigenous technologies promises a sustainable and resilient agricultural landscape. The subsequent chapters will continue to explore the dynamic intersections of tradition and innovation, unveiling the potential of Indigenous Technologies across diverse sectors.

The future of agriculture is poised for a transformative shift towards a more sustainable and resilient landscape. Several key trends and developments indicate a growing commitment to environmentally friendly practices, technological advancements, and holistic approaches that prioritize the well-being of both the planet and its inhabitants.

1. Regenerative Agriculture: Regenerative agriculture is gaining momentum as a holistic approach that goes beyond sustainable practices. It focuses on rebuilding soil health,

enhancing biodiversity, and improving the overall ecosystem. Practices such as cover cropping, agroforestry, and reduced tillage are becoming integral components of regenerative agriculture, aiming to restore and rejuvenate the land.

2. Precision Agriculture Integration: The integration of precision agriculture technologies will continue to play a pivotal role in optimizing resource use and reducing environmental impact. From precision planting and targeted pesticide application to data-driven decision support systems, farmers will increasingly leverage technology to enhance efficiency and minimize waste.

3. Climate-Smart Agriculture: As the impacts of climate change become more pronounced, the adoption of climate-smart agricultural practices will rise. This includes the development of drought-resistant crops, precision water management systems, and strategies to mitigate the effects of extreme weather events. Agriculture will become

more adaptive and resilient in the face of a changing climate.

4. Sustainable Intensification: The concept of sustainable intensification seeks to increase agricultural productivity without compromising long-term environmental sustainability. This involves optimizing inputs, improving crop yields, and reducing environmental impact through the adoption of innovative technologies, without expanding agricultural land at the expense of ecosystems.

5. Circular Agriculture and Closed-Loop Systems: A shift towards circular agriculture involves closing the loop on resource use, waste, and energy. Circular systems aim to minimize waste through practices like recycling agricultural by-products, reusing water, and creating closed-loop nutrient cycles. This approach reduces the environmental footprint of farming and promotes a more circular and sustainable agricultural model.

6. Agroecology and Organic Practices: Agroecology, which integrates ecological principles into agriculture, and organic farming practices are gaining prominence. These approaches prioritize soil health, biodiversity, and natural ecosystem services. The reduction of synthetic inputs and the promotion of biological diversity contribute to the creation of resilient and sustainable farming systems.

7. Smart Farming and AI Integration: The continued development and integration of smart farming technologies, including artificial intelligence (AI), will enable farmers to make more informed decisions. AI-driven analytics can optimize resource use, predict crop diseases, and provide real-time monitoring of farm conditions, contributing to more efficient and sustainable farming practices.

8. Global Collaboration for Sustainability: The agricultural sector is increasingly recognizing the need for global collaboration

to address sustainability challenges. Initiatives focusing on knowledge sharing, technology transfer, and collaborative research will become more prevalent, fostering a global community committed to sustainable agriculture.

9. Consumer Demand for Sustainable Practices: Growing consumer awareness and demand for sustainably produced food will influence farming practices. Farmers adopting environmentally friendly and socially responsible methods are likely to find increased market opportunities, encouraging a shift towards more sustainable and ethical production systems.

In the coming years, the convergence of these trends is expected to shape a future agricultural landscape that is not only productive but also sustainable, resilient, and mindful of the ecological balance. By embracing these principles, the agricultural sector can play a pivotal role in addressing global challenges such as climate change, biodiversity loss, and food security.

CHAPTER FIVE

Advancements in Healthcare through Indigenous Technologies

5.1 The Intersection of Tradition and Modernity in Healthcare

This chapter embarks on a journey through the evolving landscape of healthcare in India, exploring the profound impact of indigenous technologies on medical research, pharmaceuticals, and healthcare delivery. From ancient healing practices to cutting-edge innovations, the narrative highlights the diverse array of contributions that shape the well-being of the nation.

The intersection of tradition and modernity in healthcare represents a dynamic and evolving approach to wellness that draws on the strengths of both traditional and modern medical practices. This blending of ancient wisdom and cutting-edge technology seeks to provide holistic and patient-centric care, recognizing the value of diverse approaches to health and well-being.

1. Integrative Medicine: Integrative medicine combines conventional Western medicine with complementary and alternative

therapies. This approach acknowledges the benefits of traditional healing practices such as acupuncture, herbal medicine, and mind-body techniques, integrating them into mainstream healthcare to address a wide range of health issues. Integrative medicine emphasizes the importance of treating the whole person – mind, body, and spirit.

2. Herbal and Plant-Based Medicine: Traditional herbal medicine, rooted in indigenous knowledge, is experiencing a resurgence alongside the modern pharmaceutical industry. Researchers are exploring the therapeutic properties of plants and herbs, leading to the development of new drugs and treatment options. This intersection allows for the incorporation of traditional remedies into evidence-based medical practices.

3. Mind-Body Practices: Mind-body practices, such as meditation, yoga, and Tai Chi, have deep roots in traditional healing systems. These practices are now widely

embraced in modern healthcare for their proven benefits in reducing stress, improving mental health, and promoting overall well-being. Hospitals and clinics incorporate mindfulness programs and yoga classes as part of patient care.

4. Telemedicine and Digital Health: The rise of telemedicine and digital health technologies represents a modern approach to healthcare delivery. These innovations leverage the power of technology to enhance access to medical services, monitor patients remotely, and facilitate communication between healthcare providers and patients. Telemedicine respects the convenience and accessibility aspects of traditional healing by making healthcare services more widely available.

5. Traditional Diets and Nutrition: Traditional diets, often based on local and seasonal foods, are gaining recognition for their potential health benefits. Integrating traditional dietary practices into modern nutrition science allows

for a more culturally sensitive and personalized approach to nutrition. This intersection values the diversity of dietary patterns worldwide and acknowledges that there is no one-size-fits-all approach to nutrition.

6. Holistic and Preventive Healthcare: Traditional medicine often emphasizes preventive and holistic approaches to health. Modern healthcare is increasingly recognizing the importance of lifestyle factors, stress management, and preventive measures to address chronic diseases. The integration of these approaches aims to shift the focus from treating illnesses to promoting overall health and well-being.

7. Community and Cultural Competence: The intersection of tradition and modernity in healthcare emphasizes cultural competence and community involvement. Recognizing the importance of cultural beliefs, practices, and community support networks, healthcare providers strive to integrate these aspects into

patient care. This ensures that healthcare is not only effective but also culturally sensitive and respectful of diverse traditions.

8. Personalized Medicine: Modern medicine is moving towards personalized approaches that consider an individual's genetic makeup, lifestyle, and environmental factors. This aligns with the holistic principles of traditional medicine, which often tailors treatments to the unique characteristics of each patient. The integration of personalized medicine aims to provide more effective and targeted healthcare interventions.

The synergy between tradition and modernity in healthcare fosters a comprehensive and inclusive approach to well-being. It acknowledges the value of time-tested practices while leveraging advancements in medical science and technology to improve patient outcomes. This intersection encourages a collaborative and open-minded approach to healthcare that respects the diversity of healing traditions around the world.

5.2 Ayurveda: Ancient Wisdom in Modern Healthcare

The exploration begins with a tribute to Ayurveda, India's traditional system of medicine that dates back thousands of years. This section delves into the principles of Ayurveda, emphasizing its holistic approach to health and wellness. The chapter showcases how modern healthcare integrates Ayurvedic principles, leveraging herbal remedies and lifestyle recommendations for preventive and curative measures.

Ayurveda, a traditional system of medicine with roots in ancient India, has gained recognition as a holistic and natural approach to healthcare. Its principles emphasize balance and harmony within the body, mind, and spirit. In the modern context, Ayurveda is increasingly being integrated into healthcare practices worldwide.

Key Principles:

Doshas:

Ayurveda identifies three doshas - Vata, Pitta, and Kapha - representing different combinations of the five elements (earth, water, fire, air, and ether). Maintaining a balance among these doshas is crucial for overall health.

Individualized Approach:

Ayurvedic treatment is personalized, considering an individual's unique constitution (Prakriti), current imbalances (Vikriti), lifestyle, and environmental factors.

Natural Healing:

Ayurveda utilizes natural remedies derived from plants, minerals, and animals. Herbal formulations, dietary changes, and lifestyle modifications are key components of Ayurvedic healing.

Holistic Health:

Ayurveda views health as a holistic concept, encompassing physical, mental, and spiritual well-being. It focuses on preventing illnesses and promoting longevity.

Applications in Modern Healthcare:

Herbal Medicine:

Ayurvedic herbs like turmeric, ashwagandha, and neem have gained popularity for their potential health benefits. Modern research supports their anti-inflammatory, antioxidant, and immune-modulating properties.

Mind-Body Connection:

Ayurveda recognizes the interconnectedness of the mind and body. Practices such as yoga and meditation, integral to Ayurveda, are now widely acknowledged for promoting mental well-being.

Dietary Guidelines:

Ayurvedic dietary principles emphasize balancing doshas through proper food choices. Many of these principles align with current nutritional recommendations for a balanced and healthy diet.

Preventive Healthcare:

Ayurveda places a strong emphasis on preventive healthcare through lifestyle modifications, seasonal routines (Ritucharya), and detoxification practices (Panchakarma).

Stress Management:

Ayurveda provides techniques for stress reduction, including meditation, breathing exercises (Pranayama), and the use of adaptogenic herbs, addressing a critical aspect of modern healthcare.

Challenges and Considerations:

Standardization and Regulation:

Ensuring the quality and safety of Ayurvedic products and practices is a challenge. Standardization and regulatory frameworks are essential to maintain efficacy and safety.

Integration with Modern Medicine:

Collaborative efforts are needed for integrating Ayurveda into mainstream healthcare. This requires bridging the gap between traditional and modern medical systems.

Scientific Validation:

While Ayurvedic practices have a historical foundation, ongoing scientific research is necessary to validate their effectiveness and understand the mechanisms behind Ayurvedic treatments.

Future Prospects:

Global Acceptance:

Ayurveda is gaining acceptance globally, with many seeking its holistic approach. Collaborations between traditional practitioners, researchers, and the healthcare industry can contribute to its global integration.

Personalized Medicine:

Ayurveda's emphasis on individualized treatment aligns with the growing trend towards personalized medicine. Integrating Ayurvedic insights may enhance the effectiveness of modern healthcare.

Health and Wellness Tourism:

Countries with a strong Ayurvedic tradition are becoming destinations for health and wellness tourism. Ayurvedic resorts and retreats attract individuals seeking holistic rejuvenation.

In conclusion, Ayurveda, with its ancient wisdom, offers a unique perspective on healthcare that aligns with modern principles of holistic well-being. As research continues and collaborative efforts strengthen, Ayurveda's role in modern healthcare is likely to expand, providing valuable insights for a comprehensive and integrative approach to health.

5.3 Herbal Medicines and Traditional Healing Practices

As the chapter unfolds, the focus shifts to the rich tapestry of herbal medicines and traditional healing practices prevalent in different regions of India. Indigenous technologies encompass not only ancient knowledge but also innovative ways of extracting, processing, and utilizing medicinal plants. This section explores the role of traditional healers and the resurgence of interest in herbal medicine in the contemporary healthcare landscape.

1. Traditional Knowledge:

Herbal Wisdom: Traditional healing practices, rooted in cultures worldwide, have relied on the use of plants for medicinal purposes for centuries.

Indigenous Knowledge: Different cultures have developed unique herbal remedies based on local flora, passed down through generations.

2. Holistic Approach:

Balancing Body and Mind: Traditional healing often emphasizes a holistic approach, considering the interconnectedness of the body, mind, and spirit.

Preventive Care: Many traditional herbal medicines focus on preventing illnesses and maintaining overall well-being rather than just treating symptoms.

3. Diversity of Plants:

Plant-Based Remedies: Herbal medicines utilize a wide variety of plants with medicinal properties, each serving different health purposes.

Biodiversity Conservation: Traditional practices often promote sustainable harvesting and conservation of medicinal plants, contributing to biodiversity.

4. Ayurveda:

Holistic Healing System: Ayurveda, an ancient Indian system of medicine, incorporates herbal remedies, dietary guidelines, and lifestyle practices to achieve balance and harmony in the body.

Personalized Medicine: Ayurvedic treatments are often personalized, taking into account individual constitutions or doshas.

5. Scientific Validation:

Active Compounds: Many modern medicines have their roots in traditional herbal knowledge, with scientists identifying and isolating active compounds for pharmaceutical use.

Research and Studies: Ongoing scientific research validates the efficacy of certain traditional herbal remedies, leading to their incorporation into mainstream medicine.

6. Cultural Significance:

Cultural Heritage: Herbal medicines are often deeply ingrained in cultural practices and rituals, preserving cultural identity and heritage.

Community Healing: Traditional healing practices are often community-centered, fostering a sense of belonging and mutual support.

7. Challenges:

Standardization: Ensuring the quality and standardization of herbal medicines can be challenging, as formulations may vary across regions and traditions.

Integration with Modern Medicine: Bridging the gap between traditional and modern healthcare systems is a challenge, requiring collaboration and respect for both approaches.

8. Opportunities:

Complementary Medicine: Herbal medicines can complement modern healthcare, offering alternatives for certain conditions and contributing to a more holistic approach.

Economic Opportunities: The cultivation, processing, and marketing of medicinal plants can provide economic opportunities for communities, promoting sustainable practices.

9. Global Interest:

International Recognition: Traditional herbal knowledge is gaining recognition globally, with increased interest in integrating traditional practices into mainstream healthcare systems.

Cultural Exchange: Cross-cultural exchange of herbal knowledge can enrich global healthcare practices and promote a diversified approach to healing.

10. Sustainable Practices:

Conservation Practices: Integrating traditional ecological knowledge with modern conservation practices can help sustainably manage medicinal plant resources.

Ethical Harvesting: Promoting ethical harvesting practices ensures the long-term availability of medicinal plants without depleting natural habitats.

In conclusion, herbal medicines and traditional healing practices represent a rich source of healthcare knowledge that has been passed down through generations. Integrating this wisdom with modern scientific research presents opportunities for holistic healthcare approaches that address the physical, mental, and spiritual aspects of well-being while also contributing to biodiversity conservation and sustainable practices.

5.4 Integration of Traditional and Western Medicine

The narrative navigates through the integration of traditional and Western medicine in India's healthcare system. The chapter highlights examples of hospitals and healthcare practitioners combining the strengths of both systems to provide comprehensive and personalized healthcare solutions. The collaborative efforts emphasize the importance of a holistic approach that addresses physical, mental, and emotional well-being.

The integration of traditional and Western medicine, often referred to as complementary and alternative medicine (CAM) or integrative medicine, represents a holistic approach to healthcare that combines the strengths of both medical systems. This collaboration seeks to provide comprehensive and patient-centered care by recognizing the benefits of diverse healing modalities. Several aspects contribute to the integration of these two approaches:

1. Holistic Patient-Centered Care: Integrative medicine places the patient at the center of care, considering their physical, mental, emotional, and spiritual well-being. By incorporating both traditional and Western medical approaches, healthcare providers aim to address the root causes of illness and promote overall health.

2. Cross-Cultural Understanding: Healthcare professionals increasingly recognize the importance of cross-cultural understanding in patient care. Integrating traditional healing

practices acknowledges the cultural diversity of patients and fosters respectful collaboration between Western-trained healthcare providers and practitioners of traditional medicine.

3. Combined Treatment Plans: Patients may receive combined treatment plans that incorporate both conventional medical interventions and traditional healing modalities. For example, a cancer patient undergoing chemotherapy might also receive acupuncture for managing treatment-related side effects, such as nausea and fatigue.

4. Evidence-Based Practice: Integrative medicine strives to incorporate evidence-based practices from both traditional and Western medicine. This involves conducting research to assess the safety and efficacy of traditional therapies and integrating those with proven benefits into mainstream healthcare.

5. Coordinated Care Teams: Integrative healthcare often involves a collaborative and

interdisciplinary approach. Patients may have a team of healthcare professionals, including medical doctors, naturopaths, acupuncturists, nutritionists, and mental health practitioners, working together to address various aspects of their health.

6. Inclusion of Traditional Healing Practices: Traditional healing practices, such as acupuncture, herbal medicine, Ayurveda, and traditional Chinese medicine, are increasingly integrated into healthcare settings. These practices may be used to manage chronic conditions, improve well-being, and complement Western medical interventions.

7. Mind-Body Medicine: Mind-body practices, such as meditation, yoga, and mindfulness, are often integrated into treatment plans to address both physical and mental health. These practices are recognized for their therapeutic benefits in stress reduction, pain management, and overall emotional well-being.

8. Education and Training: Healthcare professionals are increasingly receiving education and training on integrative medicine approaches. This includes understanding the principles and practices of traditional healing systems to enhance their ability to collaborate with practitioners from diverse backgrounds.

9. Wellness and Preventive Medicine: The integration of traditional and Western medicine emphasizes preventive healthcare and wellness promotion. Lifestyle modifications, nutritional guidance, and stress management, often rooted in traditional practices, are incorporated to prevent illness and promote long-term health.

10. Patient Empowerment: Integrative medicine empowers patients to actively participate in their healthcare decisions. Patients are encouraged to explore and discuss complementary approaches with their healthcare team, fostering a sense of partnership and shared decision-making.

While the integration of traditional and Western medicine continues to evolve, it represents a paradigm shift towards a more inclusive, patient-centered, and holistic approach to healthcare. By embracing the strengths of both systems, this integration aims to provide more comprehensive and personalized care to individuals, recognizing the diversity of healing traditions and promoting overall well-being.

5.5 Modern Pharmaceutical Innovations

This section explores how indigenous technologies contribute to modern pharmaceutical innovations. From research and development to production processes, the narrative unveils instances where traditional knowledge inspires the creation of new drugs and therapies. The chapter showcases the synergy between indigenous wisdom and cutting-edge biotechnological advancements in the pharmaceutical industry.

In the intricate tapestry of pharmaceutical innovation, the interweaving threads of traditional knowledge and modern biotechnological advancements create a compelling narrative. This chapter explores how indigenous technologies not only contribute but also inspire novel pharmaceutical developments, showcasing a harmonious synergy between ancient wisdom and cutting-edge research in the pharmaceutical industry.

1. Ethnobotanical Discoveries: Indigenous communities have long held a treasure trove of knowledge about the medicinal properties of local plants. Researchers, recognizing the value of this traditional wisdom, delve into ethnobotanical studies to identify potential sources for new drugs. Through meticulous exploration, scientists uncover compounds with therapeutic properties that form the basis for pharmaceutical development.

2. Bioprospecting and Traditional Remedies: Bioprospecting involves the systematic search for bioactive compounds in nature, often guided by traditional remedies.

Indigenous healing practices serve as valuable guides, pointing researchers toward plants and natural substances with medicinal potential. This approach has led to the discovery of compounds used in the development of drugs for conditions ranging from pain management to cancer treatment.

3. Integrating Traditional Healing Practices: In some cases, pharmaceutical innovations draw inspiration from traditional healing practices themselves. Therapies like acupuncture, traditional Chinese medicine, or Ayurveda provide insights into holistic approaches that consider the interconnectedness of the body, mind, and spirit. Modern drug development increasingly acknowledges the importance of a holistic perspective, integrating elements of traditional healing into treatment protocols.

4. Biopiracy and Ethical Collaboration: As researchers draw upon traditional knowledge, the issue of biopiracy, or the unauthorized use of indigenous resources, becomes a concern.

Ethical collaboration becomes paramount, with pharmaceutical companies working closely with indigenous communities to ensure fair compensation, respect cultural traditions, and share benefits derived from the commercialization of traditional knowledge.

5. Traditional Formulations and Drug Development: Indigenous communities often have well-established traditional formulations for treating various ailments. Modern pharmaceutical research delves into these formulations to identify active compounds and understand their mechanisms of action. This exploration can lead to the development of new drugs or the enhancement of existing ones based on traditional formulations.

6. Biotechnological Advancements in Drug Production: Cutting-edge biotechnological tools enhance the production processes of pharmaceuticals inspired by traditional knowledge. Bioreactors, genetic engineering, and advanced fermentation techniques contribute to the efficient and scalable

production of drugs derived from natural sources, ensuring a sustainable supply chain.

7. Ethnopharmacology and Modern Drug Discovery: Ethnopharmacology, the study of traditional medicinal practices, serves as a bridge between traditional knowledge and modern drug discovery. By systematically analyzing the therapeutic properties of plants and other natural substances used in indigenous medicine, researchers identify leads for drug development, with a focus on safety and efficacy.

8. Cultural Preservation and Scientific Innovation: The collaboration between indigenous communities and the pharmaceutical industry not only fosters scientific innovation but also contributes to the preservation of cultural knowledge. Through respectful engagement, the pharmaceutical sector becomes a partner in preserving and respecting the rich tapestry of traditional healing practices.

As this chapter unfolds, it becomes evident that the integration of traditional and Western medicine in the pharmaceutical realm is a journey of mutual respect and collaboration. The synergy between indigenous wisdom and biotechnological advancements not only contributes to the development of new drugs and therapies but also promotes ethical practices and cultural preservation. In this evolving narrative, the intersection of tradition and modernity in pharmaceutical innovation creates a harmonious symphony that resonates with the healing traditions of the past and the possibilities of the future.

5.6 Telemedicine and Digital Healthcare Solutions

The chapter advances into the realm of digital healthcare, showcasing how indigenous technologies contribute to the expansion of telemedicine and other digital health solutions. From remote consultations to health monitoring apps, the narrative explores how technology bridges gaps in healthcare accessibility, especially in remote and underserved areas.

1. Remote Consultations:

Customized Solutions: Indigenous technologies contribute to the development of telemedicine platforms tailored to local needs and cultural contexts.

Multilingual Support: Incorporating indigenous languages and dialects in telemedicine interfaces ensures effective communication with diverse populations.

2. Health Monitoring Apps:

Local Health Indicators: Digital health solutions utilize indigenous knowledge to incorporate region-specific health indicators and lifestyle factors into monitoring apps.

Community Engagement: Apps designed with community input foster engagement, encouraging regular health check-ins and preventive measures.

3. Wearable Technology:

Cultural Design: Integrating wearable devices with traditional attire or accessories makes them more culturally acceptable, encouraging widespread adoption.

Preventive Health Measures: Wearables can provide real-time data on physical activity, sleep patterns, and vital signs, supporting preventive healthcare practices.

4. SMS-based Interventions:

Low-Tech Solutions: Simple SMS-based interventions leverage indigenous technologies for health education and awareness in areas with limited access to smartphones or the internet.

Community Health Messages: Tailoring messages to cultural beliefs and practices ensures better understanding and acceptance among diverse populations.

5. Indigenous Telehealth Practitioners:

Culturally Competent Care: Training indigenous healthcare practitioners in telemedicine ensures culturally competent care, understanding traditional healing practices and beliefs.

Community Trust: Indigenous telehealth practitioners often build trust within their communities, facilitating better patient engagement.

6. Collaborative Platforms:

Community Involvement: Indigenous technologies enable the creation of collaborative health platforms that involve community leaders, traditional healers, and modern healthcare professionals.

Holistic Care: Integrating traditional healing practices into digital health platforms supports a holistic approach to healthcare.

7. Drones for Healthcare Delivery:

Remote Area Access: Indigenous technologies, including drones, facilitate the delivery of medical supplies and telemedicine equipment to remote and underserved areas.

Emergency Response: Drones equipped with medical kits can provide rapid response in emergencies, improving healthcare accessibility in challenging terrains.

8. Digital Storytelling for Health Education:

Cultural Narratives: Indigenous technologies, such as digital storytelling platforms, convey health information through culturally relevant narratives.

Oral Tradition Integration: Incorporating oral traditions into digital health education materials enhances accessibility and understanding.

9. Community-Based Telehealth Hubs:

Local Empowerment: Establishing community-based telehealth hubs equipped with indigenous technologies empowers local communities to take charge of their healthcare.

Economic Opportunities: Training community members to operate telehealth hubs creates employment opportunities and enhances sustainability.

10. Data Security and Privacy:

Cultural Sensitivity: Indigenous technologies prioritize culturally sensitive approaches to data security and privacy, respecting traditional values around health information confidentiality.

Community Ownership: Involving communities in decision-making regarding data management builds trust and ensures responsible use of health data.

In conclusion, the integration of indigenous technologies into digital healthcare solutions contributes significantly to overcoming barriers in healthcare accessibility. By embracing local knowledge, cultural practices, and community engagement, these advancements not only bridge gaps but also foster a more inclusive and culturally competent approach to healthcare delivery, particularly in remote and underserved areas.

5.7 Challenges and Opportunities in Healthcare Innovation

Acknowledging the complexities of healthcare innovation, this section addresses challenges such as regulatory frameworks, ethical considerations, and the need for widespread adoption. The narrative also explores opportunities for leveraging indigenous technologies to overcome barriers and enhance healthcare outcomes for diverse populations.

Challenges:

Data Security and Privacy:

Patient Confidentiality: Protecting patient data from unauthorized access is a significant challenge in the era of digital healthcare.

Ethical Use: Balancing the benefits of data-driven healthcare with the ethical use of patient information poses ongoing challenges.

Interoperability:

Fragmented Systems: Lack of interoperability between different healthcare systems and technologies hinders seamless data exchange.

Standardization Issues: Developing and adhering to common standards is essential for effective interoperability but can be challenging to implement.

Cost and Affordability:

Initial Investment: Implementing innovative healthcare technologies often involves significant upfront costs.

Affordability for Patients: Access to advanced medical treatments and technologies may be limited by the financial constraints of patients.

Regulatory Barriers:

Approval Processes: Stringent regulatory approval processes can slow down the introduction of new healthcare innovations.

Compliance Burden: Navigating complex regulatory frameworks poses challenges for startups and small enterprises.

Resistance to Change:

Cultural Shift: Resistance from healthcare professionals and institutions to adopt new technologies and workflows.

Patient Adoption: Ensuring patient acceptance and adherence to new technologies can be challenging.

Health Inequality:

Access Disparities: Innovation may exacerbate existing healthcare inequalities if not implemented equitably.

Rural and Underserved Areas: Limited access to advanced healthcare technologies in rural and underserved areas.

Ethical Concerns:

Bias in Algorithms: Ethical challenges arise when algorithms in healthcare technologies exhibit bias, potentially leading to unfair treatment.

Informed Consent: Ensuring informed consent in the collection and use of health data is an ongoing ethical consideration.

Opportunities:

Digital Health Records:

Efficient Data Management: Electronic health records enhance data accessibility, streamline workflows, and improve patient care.

Data Analytics: Utilizing big data analytics on health records can provide valuable insights for personalized and predictive healthcare.

Telemedicine and Remote Monitoring:

Increased Access: Telemedicine bridges geographical barriers, providing healthcare access to remote and underserved populations.

Continuous Monitoring: Remote monitoring technologies enable real-time tracking of patient health, facilitating early intervention.

Artificial Intelligence (AI) and Machine Learning:

Diagnostic Capabilities: AI applications in diagnostics improve accuracy and efficiency.

Treatment Personalization: Machine learning enables personalized treatment plans based on individual patient data.

Wearable Technologies:

Preventive Health:* Wearable devices encourage proactive health management through continuous monitoring of vital signs and activity levels.

Chronic Disease Management: Wearables assist in managing chronic conditions through remote monitoring.

Genomic Medicine:

Precision Medicine: Genomic data facilitates personalized treatment plans based on an individual's genetic makeup.

Disease Prediction: Genetic information contributes to the prediction and prevention of hereditary diseases.

Blockchain Technology:

Data Security: Blockchain enhances data security by providing a decentralized and transparent system.

Interoperability: Blockchain can improve interoperability by creating a standardized and secure platform for data exchange.

Mobile Health (mHealth) Apps:

Health Education:* Mobile apps offer accessible health education and awareness programs.

Remote Consultations: mHealth apps enable virtual consultations, improving healthcare accessibility.

3D Printing in Healthcare:

Customized Solutions: 3D printing allows the creation of personalized medical devices and implants.

Supply Chain Efficiency: Localized production of medical equipment reduces dependency on centralized supply chains.

Collaborations and Partnerships:

Industry Collaboration: Collaborations between healthcare providers, tech companies, and research institutions drive innovation.

Public-Private Partnerships: Joint efforts between the public and private sectors enhance the development and implementation of healthcare innovations.

Patient Empowerment:

Health Literacy: Technology enables improved health literacy, empowering patients to make informed decisions about their health.

Active Participation: Patient engagement platforms foster active involvement in healthcare decisions and self-management.

In conclusion, the dynamic landscape of healthcare innovation presents both challenges and opportunities. Addressing the challenges requires a concerted effort from stakeholders, including regulatory bodies, healthcare providers, technology developers, and patients. Embracing opportunities can lead to transformative changes in healthcare delivery, improving accessibility, efficiency, and patient outcomes.

5.8 Future Horizons: A Healthier Nation through Indigenous Technologies

As we conclude this chapter, the focus turns towards the future of healthcare in India. The integration of indigenous technologies promises a healthcare landscape that is not only rooted in tradition but also adaptive and innovative. The subsequent chapters will continue to unveil the transformative potential of Indigenous Technologies across various sectors, contributing to the vision of a Developed India.

The future holds promising horizons for achieving a healthier nation through the integration and elevation of indigenous technologies in healthcare. This vision transcends conventional boundaries, leveraging the rich tapestry of traditional wisdom, innovative technologies, and collaborative efforts to forge a path toward holistic well-being.

1. Integrating Indigenous Healing Practices: In the coming years, healthcare systems are poised to embrace and integrate indigenous healing practices into mainstream medical care. Traditional therapies such as herbal medicine, acupuncture, and mind-body practices will be recognized for their efficacy in promoting wellness and preventing illness.

2. Personalized Medicine Rooted in Cultural Diversity: The future of healthcare will witness a shift towards personalized medicine that respects and integrates cultural diversity.

Indigenous populations often exhibit unique genetic variations, and personalized medicine will consider these differences to tailor treatments and interventions to individual needs, ensuring more effective and culturally sensitive healthcare.

3. Ethical Bioprospecting and Fair Collaboration: The ethical exploration of indigenous knowledge for drug discovery will become a cornerstone of pharmaceutical research. Collaborations between researchers, pharmaceutical companies, and indigenous communities will prioritize fair compensation, cultural respect, and shared benefits. Ethical bioprospecting will contribute to sustainable drug development while honoring traditional knowledge.

4. Technological Empowerment of Indigenous Healthcare Providers: Technological advancements will empower indigenous healthcare providers by providing access to digital health tools and telemedicine. This will enhance their capacity

to deliver healthcare services to remote or underserved areas, bridging the gap between traditional healing practices and modern medical care.

5. Indigenous Knowledge Informing Public Health Strategies: Indigenous knowledge, deeply rooted in an understanding of local ecosystems and community dynamics, will play a pivotal role in shaping public health strategies. Traditional practices that promote community well-being, preventive healthcare, and sustainable living will inform policies aimed at improving population health.

6. Culturally Tailored Mental Health Interventions: Mental health interventions will become more culturally tailored, drawing from indigenous practices that address the holistic well-being of individuals. Mindfulness techniques, traditional healing ceremonies, and community-based support systems will be integrated into mental health programs to provide more comprehensive and culturally sensitive care.

7. Preservation and Documentation of Traditional Knowledge: Efforts to preserve and document traditional knowledge will gain momentum. Collaborative initiatives between indigenous communities, researchers, and institutions will ensure the safeguarding of traditional healing practices, medicinal plant knowledge, and cultural approaches to health, fostering a sense of identity and pride.

8. Holistic Health Education: Educational programs will evolve to embrace a more holistic approach to health, incorporating elements of traditional healing practices into medical curricula. This will not only broaden the perspectives of future healthcare professionals but also foster a greater appreciation for the interconnectedness of physical, mental, and spiritual well-being.

9. Community-Driven Health Initiatives: Community-driven health initiatives, guided by indigenous values and practices, will gain prominence. Empowered communities will actively participate in designing and

implementing health programs that align with their cultural norms, ensuring sustainability and effectiveness.

As we look towards the future, the convergence of indigenous technologies and modern healthcare holds the promise of a nation where health is not merely the absence of disease but a state of holistic well-being deeply rooted in cultural traditions. By embracing the strengths of both traditional wisdom and innovative technologies, we pave the way for a healthier, more inclusive, and culturally resonant future for healthcare.

CHAPTER SIX

Challenges and Opportunities

6.1 Introduction: Navigating the Landscape of Indigenous Technologies

This chapter unravels the intricate web of challenges and opportunities embedded in the adoption and scaling of indigenous technologies. From policy landscapes to financial considerations, the narrative explores the nuances that shape the trajectory of these technologies within the dynamic ecosystem of innovation.

The adoption and scaling of indigenous technologies present a complex landscape filled with challenges and opportunities. This narrative unravels the intricate web that weaves through policy landscapes, financial considerations, and various nuances that influence the trajectory of these technologies within the dynamic ecosystem of innovation.

1. Policy Landscapes and Regulatory Frameworks: One of the primary challenges lies in navigating diverse policy landscapes and regulatory frameworks. Indigenous technologies often emerge from traditional

knowledge that may not neatly fit within existing regulations. Crafting policies that strike a balance between protecting intellectual property, ensuring cultural respect, and promoting innovation becomes crucial for the successful adoption of these technologies.

2. Cultural Sensitivity and Community Engagement: The adoption of indigenous technologies requires a deep understanding of cultural sensitivities and active engagement with the communities that hold this knowledge. Respectful collaboration is essential to avoid exploitation and ensure that the benefits of technology adoption are shared equitably with indigenous communities.

3. Financial Considerations and Funding Mechanisms: Financial barriers often impede the development and scaling of indigenous technologies. Lack of funding, limited access to resources, and the high costs associated with research and development pose

significant challenges. Establishing funding mechanisms that support the development of these technologies while respecting indigenous rights and values is critical for sustained progress.

4. Intellectual Property Rights and Benefit-Sharing: Protecting intellectual property rights while respecting traditional knowledge is a delicate balance. The adoption of indigenous technologies requires frameworks that acknowledge and protect the intellectual contributions of indigenous communities. Additionally, benefit-sharing mechanisms must be established to ensure that communities receive fair compensation for the commercial use of their knowledge.

5. Bridging Traditional and Modern Knowledge Systems: Harmonizing traditional and modern knowledge systems is a nuanced challenge. Bridging the gap between indigenous technologies and mainstream scientific methodologies requires cross-cultural collaboration, mutual understanding,

and recognition of the validity of different knowledge systems. Establishing hybrid models that integrate the strengths of both systems is crucial for successful adoption.

6. Education and Awareness Building: Education and awareness building are pivotal in fostering an environment that supports the adoption of indigenous technologies. This involves not only educating policymakers, researchers, and industry stakeholders but also raising awareness within indigenous communities about the potential benefits and risks associated with the adoption of new technologies.

7. Infrastructure and Access to Technology: Limited infrastructure and access to technology in remote or underserved areas pose significant challenges. Ensuring that indigenous communities have the necessary infrastructure and access to technological tools is essential for the effective adoption and scaling of indigenous technologies,

promoting inclusivity in the innovation ecosystem.

8. Sustainable Development Goals and Indigenous Technologies: Aligning the adoption of indigenous technologies with sustainable development goals is an opportunity to address pressing global challenges. Indigenous technologies often offer sustainable solutions for issues such as environmental conservation, healthcare, and food security. Aligning these technologies with broader development objectives can attract support and foster collaboration.

9. Knowledge Transfer and Capacity Building: Facilitating knowledge transfer and capacity building is vital for the successful adoption of indigenous technologies. This involves empowering indigenous communities with the skills and resources needed to actively participate in the development, implementation, and scaling of these technologies.

In navigating the adoption and scaling of indigenous technologies, recognizing the interconnected challenges and opportunities is essential. Crafting inclusive policies, respecting cultural values, addressing financial considerations, and fostering collaborative approaches are key elements in ensuring that the adoption of these technologies contributes positively to both indigenous communities and the broader landscape of innovation.

6.2 Policy Support: The Pillar of Sustainable Adoption

Policy frameworks play a pivotal role in facilitating or hindering the integration of indigenous technologies. This section dissects the existing policies, highlighting their impact on innovation, research, and development. The chapter delves into the need for policy reforms that foster an environment conducive to the growth of indigenous technologies across various sectors.

In the dynamic landscape of innovation, the role of policy frameworks is both intricate and decisive. The integration and scaling of indigenous technologies face a series of challenges and opportunities shaped by the prevailing policy environment. This section scrutinizes existing policies, shedding light on their impact on innovation, research, and development within the context of indigenous technologies. Moreover, it delves into the imperative for policy reforms that can cultivate an environment conducive to the expansive growth of these technologies across diverse sectors.

1. Impact of Current Policies on Innovation: The narrative begins by unraveling the impact of existing policies on the innovation landscape. Policies governing intellectual property rights, research funding, and technology transfer are scrutinized to discern their influence on the development and proliferation of indigenous technologies. It explores whether current frameworks

encourage or hinder the exploration and utilization of traditional knowledge and technologies.

2. Research and Development Initiatives: An in-depth analysis of policies governing research and development initiatives reveals their role in shaping the trajectory of indigenous technologies. The section explores how funding mechanisms, research priorities, and collaborative frameworks influence the discovery, validation, and application of indigenous knowledge in technological advancements. It also investigates the inclusion of traditional knowledge holders in research processes.

3. Need for Policy Reforms: Delving into the intricacies of the policy landscape, the chapter emphasizes the imperative for policy reforms. It explores the necessity of crafting policies that not only recognize the value of indigenous technologies but actively support their integration into mainstream innovation ecosystems. This includes addressing legal

and regulatory frameworks to ensure the protection of traditional knowledge while fostering innovation.

4. Indigenous Technology Commercialization: The commercialization of indigenous technologies requires policies that facilitate market access, protect intellectual property, and ensure fair compensation for indigenous communities. The section dissects existing policies related to technology commercialization, examining whether they provide an enabling environment for indigenous entrepreneurs and innovators to bring their technologies to market.

5. Cultural Heritage Protection: Beyond technological considerations, the chapter explores policies related to the protection of cultural heritage. It examines how policies address the ethical use of indigenous knowledge, ensuring that technological innovations respect cultural values and traditions. The discussion includes

considerations of informed consent, benefit-sharing, and the safeguarding of cultural practices.

6. Inclusivity in Policy Formulation: An essential aspect of fostering an environment conducive to indigenous technologies is ensuring inclusivity in policy formulation. The chapter explores the importance of engaging indigenous communities, traditional knowledge holders, and relevant stakeholders in the policymaking process. Inclusive policies are more likely to capture the nuances of indigenous technologies and address the unique challenges they face.

7. International Collaboration and Harmonization: Given the global nature of innovation, the narrative delves into the role of international collaboration and harmonization of policies. It explores how policies at the national and international levels can align to support the adoption and scaling of indigenous technologies,

facilitating cross-border collaborations and knowledge exchange.

8. Overcoming Financial Barriers: Financial considerations are a crucial aspect of the adoption and scaling of indigenous technologies. The section examines existing financial policies and mechanisms, exploring whether they adequately support indigenous innovators and entrepreneurs. It also discusses the need for financial reforms that ensure equitable access to resources for those working with indigenous technologies.

As the narrative unfolds, it becomes evident that the adoption and scaling of indigenous technologies are intricately tied to the policies that govern innovation. The chapter not only dissects the challenges posed by current policies but also underscores the opportunities for reform. By navigating these policy landscapes with a nuanced understanding of indigenous technologies, policymakers can pave the way for an inclusive and innovation-driven future.

6.3 Financial Investments: Nurturing Innovation

Financial backing is a linchpin in the journey from concept to execution for indigenous technologies. The narrative explores the challenges faced by innovators in securing funding, be it through government grants, private investments, or venture capital. The chapter also sheds light on successful funding models and the potential for creating investment ecosystems that support indigenous technological ventures.

Securing funding for innovative projects can be a challenging yet crucial aspect of bringing new technologies to market. The challenges faced by innovators in securing funding can be attributed to various factors, and the sources of funding, such as government grants, private investments, and venture capital, each present their own set of obstacles.

Government Grants:

Competitive Nature: Government grants are often highly competitive, with a limited pool of funds available. Innovators need to demonstrate the uniqueness and potential impact of their projects to stand out in the application process.

Bureaucratic Processes: Securing government funding can involve complex application processes and bureaucratic hurdles, leading to delays and uncertainties for innovators.

Private Investments:

Risk Aversion: Private investors are often risk-averse and may hesitate to invest in innovative projects with uncertain outcomes. Convincing investors of the potential success and return on investment can be a significant challenge.

Lack of Tangible Results: Innovators may face difficulties attracting private investments if their projects are at an early stage and lack tangible results or market validation.

Venture Capital:

Market Readiness: Venture capitalists typically seek projects that are market-ready or have a proven track record. Early-stage innovators may find it challenging to secure venture capital without a solid business plan and market traction.

Equity Considerations: Venture capital often involves giving up a portion of ownership in the company, which might not be favorable for some innovators. Negotiating fair terms while meeting the needs of both parties can be complex.

Successful Funding Models:

Collaborative Funding: Successful innovators often leverage a combination of funding sources, creating a diverse and resilient financial foundation. This may include a mix of government grants, private investments, and venture capital.

Strategic Partnerships: Forming strategic partnerships with established companies or industry players can provide not only funding but also access to resources, expertise, and market channels.

Creating Investment Ecosystems:

Incubators and Accelerators: Developing and supporting incubators and accelerators that nurture and guide innovative projects can create an ecosystem where investors are more willing to engage with emerging technologies.

Networking Events and Platforms: Establishing events, conferences, and online platforms that connect innovators with potential investors fosters collaboration and increases the visibility of promising projects.

Policy Support: Governments can play a crucial role by implementing policies that encourage investment in innovation, providing tax incentives, and reducing regulatory barriers.

In conclusion, overcoming funding challenges requires a combination of strategic planning, effective communication, and adaptability. Successful funding models often involve a mix of different sources, and creating a supportive investment ecosystem is essential for the sustained growth of indigenous technological ventures.

6.4 Overcoming Technological Barriers

Innovation often faces technological barriers, ranging from infrastructural challenges to limited access to cutting-edge tools. This section explores strategies to overcome these hurdles, emphasizing the importance of creating a robust technological infrastructure and providing access to resources that enable researchers and innovators to thrive.

Indeed, innovation frequently encounters various technological barriers that can impede progress. These barriers encompass a wide range of challenges, from inadequate infrastructure to limited access to state-of-the-art tools and technologies. Here are some common technological barriers that innovators may face:

Infrastructural Challenges:

Limited Connectivity: Insufficient or unreliable internet connectivity can hinder the exchange of information and collaboration among innovators. This is especially critical for projects that require real-time data sharing or cloud-based resources.

Power Supply Issues: In regions with inconsistent or unreliable power supply, innovators may struggle to maintain continuous operations, particularly if their projects rely heavily on electricity.

Access to Cutting-edge Tools and Technologies:

High Costs: Acquiring and using cutting-edge tools and technologies can be expensive, making it challenging for innovators, especially those with limited financial resources, to access the latest advancements in their field.

Technological Obsolescence: Rapid advancements in technology can lead to the obsolescence of existing tools, creating a barrier for innovators who must continually invest in updating their equipment.

Lack of Research and Development Infrastructure:

Limited R&D Facilities: Insufficient research and development facilities can hinder the ability of innovators to conduct experiments, test prototypes, and refine their ideas. This is particularly relevant in industries where physical testing and experimentation are crucial.

Intellectual Property Challenges:

Inadequate IP Protection: Weak or inadequate intellectual property protection can discourage innovators from investing in the development of new technologies, as they may fear that their innovations could be easily replicated or exploited by others.

Skills and Expertise Gap:

Lack of Trained Personnel: Innovators may face challenges in finding skilled professionals with the necessary expertise to work on cutting-edge projects. This skills gap can slow down the pace of innovation.

Interoperability Issues:

Lack of Standardization: In industries where interoperability is crucial, the absence of standardized protocols and technologies can impede the seamless integration of different systems, hindering innovation.

Data Security and Privacy Concerns:

Cybersecurity Risks: As innovation often involves the use of digital technologies, concerns about data security and privacy can be significant barriers. Innovators must navigate the complexities of ensuring the safety and confidentiality of sensitive information.

Addressing these technological barriers requires a multi-faceted approach involving collaboration between government, industry, and research institutions. Investments in infrastructure, education, and research and development can help create an environment where innovators have the tools and resources needed to overcome technological challenges and drive meaningful advancements in their respective fields.

6.5 Public Awareness: Bridging the Gap

The success of indigenous technologies is intricately linked to public acceptance and awareness. This section examines the challenges in communicating the value of these technologies to the public and highlights strategies for bridging the awareness gap. It also explores the role of education and outreach programs in fostering a culture that embraces indigenous innovations.

Communicating the value of innovative technologies to the public is a crucial aspect of ensuring their acceptance and adoption. The challenges in conveying this value often stem from the complexity of the technologies, skepticism, or a lack of awareness. Bridging the awareness gap involves strategic communication, education, and outreach efforts. Here's an exploration of the challenges and strategies:

Challenges in Communicating the Value of Technologies:

Complexity and Technical Jargon:

Challenge: Innovative technologies are often complex and involve technical jargon that may be difficult for the general public to understand.

Strategy: Simplify the language and use relatable analogies to convey the benefits and applications of the technology. Communicate in a way that resonates with the target audience.

Skepticism and Fear of the Unknown:

Challenge: The public may be skeptical or fearful of new technologies, especially if they are perceived as disruptive or if there are concerns about safety and ethical implications.

Strategy: Engage in transparent communication, addressing concerns proactively. Provide clear information about safety measures, ethical considerations, and potential positive impacts on society.

Limited Access and Exposure:

Challenge: Some segments of the population may have limited access to information or exposure to innovative technologies.

Strategy: Implement outreach programs that target diverse communities, ensuring that information about the technologies reaches all demographics. Utilize various communication channels, including social media, community events, and traditional media.

Cultural and Social Barriers:

Challenge: Cultural biases and social norms can affect the perception of certain technologies, leading to resistance or misunderstanding.

Strategy: Tailor communication strategies to address cultural sensitivities. Collaborate with community leaders and influencers to convey the cultural relevance and benefits of the technologies.

Strategies for Bridging the Awareness Gap:

Public Awareness Campaigns:

Develop comprehensive and engaging public awareness campaigns that highlight the positive impact of the technologies on individuals and society. Utilize multimedia formats to reach a wider audience.

Stakeholder Engagement:

Involve various stakeholders, including community leaders, policymakers, and influencers, in the communication process. Their endorsement and support can enhance the credibility of the technologies.

Education and Outreach Programs:

Implement educational initiatives to enhance public understanding of the technologies.

Work with schools, colleges, and community organizations to integrate technology-related education into curricula and outreach programs.

Interactive Platforms and Demonstrations:

Organize interactive events, workshops, and demonstrations that allow the public to experience the technologies firsthand. This hands-on approach can demystify complex concepts and create a positive impression.

Inclusive Communication:

Ensure that communication materials are inclusive and accessible to diverse audiences. Consider language, visuals, and cultural nuances to make the information relatable to different demographic groups.

Role of Education and Outreach Programs:

Cultivating Curiosity and Interest:

Develop education programs that foster curiosity and interest in science and technology from an early age. Encourage hands-on learning experiences and extracurricular activities that showcase the practical applications of technology.

Community Involvement:

Engage communities in the co-creation and development of technology projects. Community involvement fosters a sense of ownership and pride, leading to increased acceptance of indigenous innovations.

Lifelong Learning Initiatives:

Establish lifelong learning initiatives that provide continuous education and updates on technological advancements. This helps the public stay informed and adapt to evolving technologies.

Partnerships with Educational Institutions:

Collaborate with schools, colleges, and universities to integrate information about indigenous innovations into educational curricula. This ensures that future generations are well-informed about and supportive of these technologies.

In conclusion, effective communication and education efforts play a pivotal role in bridging the awareness gap and fostering a culture that embraces indigenous innovations. By addressing challenges head-on and implementing targeted strategies, innovators can build trust, inspire curiosity, and cultivate public support for the positive contributions of these technologies to society.

6.6 Opportunities in Indigenous Technologies

Amidst the challenges, the chapter shifts focus to the myriad of opportunities that arise from the embrace of indigenous technologies. It explores how these technologies can spur economic growth, create employment opportunities, and contribute to sustainable development. The narrative also delves into the potential for global collaborations and market expansion.

Embracing indigenous technologies presents a myriad of opportunities that extend beyond technological advancement. This shift in focus recognizes the potential for indigenous technologies to drive economic growth, create employment opportunities, and contribute to sustainable development. Additionally, it explores the possibilities for global collaborations and market expansion. Here's an exploration of these opportunities:

1. Economic Growth:

Innovation Ecosystems: Embracing indigenous technologies fosters the development of innovation ecosystems. These ecosystems, consisting of startups, research institutions, and established businesses, contribute to economic growth by generating new products, services, and business models.

Entrepreneurship: Indigenous technologies often lead to the emergence of new entrepreneurs who create innovative solutions. This entrepreneurial activity stimulates economic development by generating revenue, attracting investments, and fostering a culture of innovation.

2. Employment Opportunities:

Skill Development: The adoption of indigenous technologies necessitates skilled labor, leading to increased demand for technical expertise. This, in turn, creates opportunities for skill development programs and vocational training, contributing to higher employment rates.

Job Creation: Startups and companies involved in the development and implementation of indigenous technologies create job opportunities across various sectors, including research and development, manufacturing, marketing, and support services.

3. Sustainable Development:

Environmental Impact: Indigenous technologies often prioritize sustainability and environmental considerations. Green technologies, renewable energy solutions, and eco-friendly practices contribute to sustainable development by reducing the environmental footprint of various industries.

Inclusive Development: Indigenous technologies can be tailored to address local challenges, promoting inclusive development. Solutions that improve access to healthcare, education, and essential services contribute to a more equitable and sustainable society.

4. Global Collaborations:

Knowledge Exchange: Embracing indigenous technologies opens avenues for global knowledge exchange. Collaborations with international partners, research institutions, and businesses facilitate the sharing of expertise, best practices, and technological advancements.

Joint Research and Development: Global collaborations enable joint research and development initiatives, leading to the creation of more robust and globally competitive technologies. Shared resources and diverse perspectives enhance the quality and impact of innovations.

5. Market Expansion:

Export of Technologies: Successful indigenous technologies can be exported to international markets, contributing to economic gains. This expansion into global markets enhances the visibility and reputation of indigenous innovations.

International Investments: The acceptance of indigenous technologies can attract international investments. Foreign investors may recognize the potential for growth and innovation, leading to partnerships and financial support for further development.

6. Social Impact:

Improving Quality of Life: Indigenous technologies, particularly those focused on healthcare, education, and social services, have the potential to significantly improve the quality of life for communities. This social impact contributes to the overall well-being of society.

Community Empowerment: Embracing indigenous technologies empowers local communities by providing them with tools and solutions to address their unique challenges. This empowerment contributes to social resilience and self-sufficiency.

In conclusion, the embrace of indigenous technologies holds immense potential for multifaceted opportunities. Beyond technological advancement, these opportunities extend to economic growth, job creation, sustainable development, global collaborations, and market expansion. By recognizing and capitalizing on these prospects, societies can harness the full benefits of indigenous innovations, fostering a positive and transformative impact on a local and global scale.

6.7 Fostering a Culture of Innovation and Self-Reliance

The chapter concludes by emphasizing the need to cultivate a culture of innovation and self-reliance. It explores how educational institutions, research centers, and the industry can work collaboratively to nurture a mindset that values indigenous technologies. The chapter leaves readers with a vision of a future where innovation becomes ingrained in the nation's DNA, driving continuous progress.

Cultivating a culture of innovation and self-reliance is essential for sustained progress and the widespread adoption of indigenous technologies. This chapter highlights the importance of collaboration among educational institutions, research centers, and industry to nurture a mindset that values innovation. It envisions a future where innovation becomes ingrained in the nation's DNA, driving continuous progress. Here's an exploration of these key themes:

1. Educational Institutions:

Curriculum Enhancement: Educational institutions can play a pivotal role in fostering a culture of innovation by integrating innovation-focused modules and practical experiences into their curricula. This includes hands-on projects, case studies, and collaborations with industry partners.

Entrepreneurial Programs: Developing entrepreneurship programs within educational institutions encourages students to apply their knowledge to real-world problems, fostering a mindset of innovation and self-reliance.

2. Research Centers:

Applied Research: Research centers can focus on applied research that addresses practical challenges and opportunities. This approach ensures that innovations are not confined to theoretical concepts but are relevant and applicable to real-world scenarios.

Collaboration with Industry: Establishing strong ties with industry partners enables research centers to align their efforts with market needs. Collaborative projects can accelerate the development and implementation of indigenous technologies.

3. Industry Collaboration:

Open Innovation Practices: Industries can adopt open innovation practices, collaborating with external partners, startups, and research institutions. This facilitates the exchange of ideas, resources, and expertise, fostering a culture of continuous improvement and innovation.

Investment in R&D: Industries should prioritize investments in research and development (R&D) to drive innovation within their operations. This commitment to R&D not only leads to technological advancements but also inspires a culture of curiosity and exploration.

4. Collaborative Initiatives:

Innovation Hubs and Clusters: Establishing innovation hubs and clusters that bring together academia, research, and industry creates a collaborative ecosystem. These hubs serve as focal points for knowledge exchange, idea generation, and project development.

Joint Training Programs: Collaborative initiatives can include joint training programs where industry professionals, researchers, and students participate in workshops and seminars. This interdisciplinary approach fosters a holistic understanding of innovation.

5. Vision for the Future:

Ingraining Innovation in Society: The ultimate vision is to ingrain innovation in the nation's DNA, where creativity, problem-solving, and continuous improvement become integral to the societal mindset.

Cultural Transformation: Envisioning a future where innovation is embraced as a cultural norm involves breaking down resistance to change, encouraging risk-taking, and celebrating both successes and failures as learning experiences.

6. Importance of Self-Reliance:

Reducing Dependency: Cultivating self-reliance involves reducing dependency on external technologies and solutions.

Indigenous technologies tailored to local needs enhance resilience and sustainability.

Empowering Communities: Self-reliance empowers communities to address their unique challenges, fostering a sense of ownership and pride in their ability to create solutions.

As we move forward, the subsequent chapters will continue to build upon the foundation laid in this exploration of challenges and opportunities. The focus will shift to specific sectors and their unique dynamics within the landscape of Indigenous Technologies.

CHAPTER SEVEN

Future Prospects

7.1 Navigating the Technological Horizon

In this concluding chapter, we cast our gaze towards the future, envisioning the role and trajectory of indigenous technologies in shaping the destiny of India. The narrative unfolds into the promising landscape of emerging fields and the continued growth of traditional knowledge systems, offering a glimpse into a future driven by innovation and technological prowess.

As we cast our gaze towards the future, the trajectory of indigenous technologies in shaping the destiny of India is a promising and transformative journey. The narrative unfolds into a landscape where emerging fields and the continued growth of traditional knowledge systems converge, offering a glimpse into a future driven by innovation and technological prowess.

1. Emerging Fields:

Artificial Intelligence (AI) and Machine Learning (ML): Indigenous technologies in AI and ML are poised to play a pivotal role in revolutionizing sectors such as healthcare, agriculture, and finance. From personalized medicine to precision agriculture, India's innovations in AI and ML will contribute to efficiency and effectiveness.

Renewable Energy: India, with its focus on sustainability, is likely to lead in the development and implementation of indigenous technologies in renewable energy. Innovations in solar, wind, and other clean energy sources will propel the nation towards energy self-sufficiency and reduced environmental impact.

Biotechnology and Healthcare: Indigenous technologies in biotechnology are anticipated to drive breakthroughs in healthcare, ranging from personalized medicine to advanced diagnostic tools. These innovations will contribute to improved health outcomes and a robust healthcare ecosystem.

2. Growth of Traditional Knowledge Systems:

Integration with Modern Technologies: The future will witness a seamless integration of traditional knowledge systems with modern technologies. From Ayurveda-inspired healthcare solutions to sustainable agricultural practices rooted in ancient wisdom, the synergy of tradition and technology will pave the way for holistic development.

Preservation of Biodiversity: Indigenous technologies will play a crucial role in the preservation of biodiversity. Traditional agricultural practices, coupled with modern advancements, will contribute to sustainable farming methods that maintain ecological balance and biodiversity conservation.

Cultural Heritage Preservation: Technology will be harnessed to preserve and promote India's rich cultural heritage. Virtual reality, augmented reality, and other immersive technologies will offer innovative ways to showcase and pass on traditional art, music, and cultural practices to future generations.

3. Innovation Hubs and Collaboration:

Global Leadership in Innovation: India is poised to become a global leader in innovation, with dedicated innovation hubs and collaborative ecosystems. These hubs will bring together diverse talents, foster interdisciplinary collaborations, and attract international attention for cutting-edge advancements.

Cross-Sector Collaborations: The future will see increased collaborations across sectors, with academia, industry, and government working hand in hand. This collaborative approach will accelerate the development and deployment of indigenous technologies, addressing societal challenges and fostering economic growth.

4. Socio-Economic Impact:

Employment Opportunities: Indigenous technologies will be a driving force behind job creation, especially in emerging fields. The growth of startups, technology-driven enterprises, and a skilled workforce will contribute to reducing unemployment and fostering economic prosperity.

Inclusive Development: The benefits of indigenous technologies will reach all strata of society, promoting inclusive development. From smart urban planning to rural empowerment through technology, the future envisions a balanced and equitable socio-economic landscape.

5. Technological Diplomacy:

Global Collaboration and Market Expansion: India's indigenous technologies will not only contribute to national development but also position the country as a key player in global technological advancements. Technological diplomacy will open avenues for collaboration, trade, and international recognition.

Soft Power Projection: Technological prowess will become a significant component of India's soft power. The positive impact of indigenous technologies on global challenges, such as climate change and healthcare, will enhance India's global standing and influence.

In conclusion, the future of India is shaped by the dynamic interplay of indigenous technologies, emerging fields, and the preservation of traditional knowledge systems. As the nation continues to innovate and embrace technological prowess, it envisions a future where the synergy of tradition and modernity propels India towards sustainable development, global leadership, and a legacy of transformative impact on the world stage.

7.2 Emerging Frontiers: Artificial Intelligence and Biotechnology

As we step into the future, the integration of cutting-edge technologies becomes paramount. This section explores the potential of indigenous technologies in emerging fields, with a special focus on artificial intelligence and biotechnology. From AI-driven solutions to genetic advancements, the chapter delves into how these technologies can revolutionize industries and contribute to the overall development of the nation.

As the integration of cutting-edge technologies becomes paramount, the potential of indigenous technologies in emerging fields takes center stage. This section explores the transformative impact of indigenous technologies, with a special focus on two pivotal areas: Artificial Intelligence (AI) and Biotechnology. From AI-driven solutions to genetic advancements, the chapter delves into how these technologies can revolutionize industries and contribute to the overall development of the nation.

1. Artificial Intelligence (AI):

AI in Healthcare: Indigenous AI technologies are poised to revolutionize healthcare. AI-driven diagnostics, personalized treatment plans, and predictive analytics will enhance patient care, improve medical outcomes, and make healthcare more accessible and efficient.

Smart Agriculture: AI applications in agriculture, tailored to Indian conditions, can optimize crop management, predict disease outbreaks, and improve yield. Precision farming techniques driven by AI will contribute to sustainable agriculture and food security.

Financial Technology (Fintech): Indigenous AI solutions in the financial sector will lead to innovations in fraud detection, risk assessment, and personalized financial services. These advancements will foster financial inclusion and enhance the efficiency of the banking and financial systems.

Smart Cities: AI technologies will play a crucial role in the development of smart cities. From intelligent traffic management to predictive maintenance of infrastructure, indigenous AI applications will create more sustainable, efficient, and livable urban environments.

Education and E-Learning: AI-driven personalized learning platforms will transform the education landscape. Adaptive learning systems, virtual tutors, and data-driven insights will cater to diverse learning needs, fostering a knowledge-driven society.

2. Biotechnology:

Genomic Medicine: Indigenous biotechnological advancements will pave the way for personalized genomic medicine. Genetic diagnostics, gene therapies, and precision medicine will revolutionize healthcare by tailoring treatments to individual genetic profiles.

Agricultural Biotechnology: Genetically modified crops and innovative biotechnological solutions will enhance crop resilience, nutrient content, and resistance to pests. These advancements will contribute to food security and sustainable agricultural practices.

Environmental Biotechnology: Indigenous technologies in environmental biotechnology will address pollution and waste management challenges. Bioremediation, bioenergy production, and sustainable waste treatment methods will contribute to a cleaner and healthier environment.

Biopharmaceuticals: The development of indigenous biopharmaceuticals will reduce dependency on imported drugs. Biotechnological research will lead to the production of vaccines, therapeutic proteins, and other biologics, contributing to public health.

Industrial Biotechnology: Biotechnological processes for industrial applications, such as bio-based materials and biofuels, will foster a more sustainable approach to manufacturing and contribute to the circular economy.

3. Cross-Cutting Applications:

Interdisciplinary Collaborations: The convergence of AI and biotechnology will give rise to innovative solutions with cross-cutting applications. For example, AI-driven analysis of biological data can accelerate drug discovery and optimize treatment protocols in healthcare.

Ethical Considerations: As these technologies advance, ethical considerations around data privacy, genetic manipulation, and the responsible use of AI become paramount. Indigenous technologies will need to incorporate ethical frameworks to ensure societal well-being.

4. National Development Impact:

Economic Growth: The integration of indigenous technologies in AI and biotechnology will contribute significantly to economic growth. Startups and established companies in these fields will drive innovation, create employment opportunities, and attract investments.

Global Competitiveness: A strong focus on indigenous technologies will enhance India's global competitiveness in AI and biotechnology. The nation's contributions to cutting-edge research and solutions will position it as a key player on the global technological stage.

Societal Well-Being: The application of AI and biotechnology in healthcare, agriculture, and other sectors will directly impact societal well-being. Improved healthcare outcomes, enhanced food security, and sustainable practices will contribute to an improved quality of life for the population.

In conclusion, the integration of cutting-edge technologies, particularly in AI and biotechnology, holds immense promise for the development of the nation. Indigenous advancements in these fields have the potential to reshape industries, address societal challenges, and position India as a leader in the global technological landscape. The chapter envisions a future where the strategic deployment of AI and biotechnology leads to sustainable development, economic prosperity, and a higher quality of life for the citizens of the nation.

7.3 The Resurgence of Traditional Knowledge Systems

While embracing modern technologies, the chapter emphasizes the importance of preserving and promoting traditional knowledge systems. By tapping into the wealth of wisdom passed down through generations, India can harness the strengths of traditional practices in agriculture, healthcare, and more. This section explores the symbiotic relationship between ancient wisdom and contemporary innovation.

Emphasizing the importance of preserving and promoting traditional knowledge systems, this section underscores how tapping into the wealth of wisdom passed down through generations can empower India to harness the strengths of traditional practices. The exploration focuses on the symbiotic relationship between ancient wisdom and contemporary innovation, particularly in areas such as agriculture, healthcare, and more.

1. Agriculture and Traditional Farming Practices:

Biodiversity Conservation: Traditional agricultural practices, deeply rooted in local ecosystems, promote biodiversity. Crop rotation, mixed cropping, and agroforestry techniques are time-tested methods that contribute to soil health, pest management, and sustainable farming.

Water Conservation: Indigenous methods of water conservation, such as traditional rainwater harvesting and aqueduct systems, offer sustainable solutions to address water scarcity in agriculture. These practices can be integrated with modern technologies for optimal results.

Organic Farming: Traditional farming often involves organic and natural cultivation methods. By incorporating these practices, India can contribute to global efforts in promoting sustainable and environmentally friendly agricultural practices.

2. Healthcare and Traditional Medicine:

Ayurveda and Herbal Medicine: India's traditional knowledge in Ayurveda emphasizes holistic healthcare using herbal medicines and natural remedies. Integrating Ayurvedic principles with modern medical practices can lead to personalized and preventive healthcare solutions.

Yoga and Wellness: Traditional practices like yoga, meditation, and mindfulness have gained global recognition for their positive impact on mental and physical well-being. Promoting these practices aligns with contemporary approaches to stress management and overall wellness.

Ethnobotany: Traditional knowledge of local plants and their medicinal properties, often passed down through generations, can be leveraged for drug discovery and the development of pharmaceuticals. Ethnobotanical research can bridge the gap between traditional wisdom and modern medicine.

3. Sustainable Living Practices:

Traditional Architecture: Indigenous architectural practices, such as vernacular architecture, showcase sustainable building techniques that consider local climate conditions. Integrating these practices can contribute to energy-efficient and eco-friendly construction.

Waste Management: Traditional waste management practices, including recycling and repurposing, align with contemporary sustainability goals. By incorporating these methods into modern waste management systems, India can address environmental challenges.

Community-Based Conservation: Traditional communities often engage in conservation practices that ensure the sustainable use of natural resources. Collaborating with indigenous communities for ecosystem conservation aligns with contemporary conservation efforts.

4. Symbiotic Relationship with Innovation:

Hybrid Solutions: The symbiotic relationship between traditional knowledge and contemporary innovation involves creating hybrid solutions that integrate the strengths of both. This approach leads to innovative practices that are culturally sensitive and contextually relevant.

Community Engagement: Involving local communities in the development and implementation of solutions ensures that traditional knowledge is respected and preserved. Community engagement fosters a sense of ownership and sustainability in the application of traditional practices.

Digital Platforms for Knowledge Preservation: Utilizing digital platforms to document and preserve traditional knowledge systems ensures their accessibility for future generations. Digital databases, mobile applications, and online resources can facilitate the dissemination of traditional wisdom.

5. Cultural Heritage and Identity:

Preservation of Cultural Heritage: Embracing traditional knowledge is not only about practical applications but also about preserving cultural heritage and identity. It reinforces a sense of pride and continuity with ancestral practices.

Inter-generational Knowledge Transfer: Efforts to promote traditional knowledge involve inter-generational knowledge transfer. Initiatives that encourage the passing down of skills and wisdom from elders to the younger generation contribute to the preservation of cultural identity.

In conclusion, the symbiotic relationship between ancient wisdom and contemporary innovation presents a unique opportunity for India to address current challenges while preserving its rich cultural heritage. By tapping into traditional knowledge systems in agriculture, healthcare, and sustainable living practices, India can create a harmonious blend of tradition and innovation, leading to holistic and culturally resonant solutions for the well-being of its people and the sustainable development of the nation.

7.4 Sustainable Development through Indigenous Technologies

Sustainability becomes a cornerstone in the vision for the future. The narrative explores how indigenous technologies can contribute to sustainable development, addressing environmental challenges, and fostering eco-friendly practices. From renewable energy solutions to waste management innovations, the chapter envisions a future where indigenous technologies play a pivotal role in achieving balance and harmony with nature.

1. Renewable Energy Solutions:

Solar Power Innovations: Indigenous technologies in solar power, including advancements in photovoltaic systems and solar water heaters, contribute to clean energy generation. These innovations reduce dependence on fossil fuels, mitigating climate change impacts.

Wind Energy: Indigenous wind energy solutions, such as improved turbine designs and efficient power transmission methods, harness the power of the wind for electricity generation. Wind farms can be strategically implemented to maximize energy production while minimizing environmental impact.

Hydropower: Innovative approaches to hydropower, including small-scale hydroelectric projects and river current generators, provide sustainable energy solutions. Indigenous technologies ensure that hydropower projects are designed with minimal ecological disruption.

2. Sustainable Agriculture Practices:

Agroecological Techniques: Indigenous agricultural practices, passed down through generations, emphasize agroecological techniques such as polyculture, agroforestry, and organic farming. These methods enhance soil fertility, reduce reliance on synthetic inputs, and promote biodiversity.

Precision Farming: Integrating indigenous knowledge with modern precision farming technologies optimizes resource use. Precision agriculture, including data-driven decision-making and sensor technologies, enhances crop yield while minimizing environmental impact.

Water Conservation: Indigenous water conservation practices, such as traditional rainwater harvesting and community-managed irrigation systems, contribute to sustainable agriculture. Combining these practices with modern irrigation technologies ensures efficient water use.

3. Waste Management Innovations:

Circular Economy Practices: Indigenous waste management practices often align with circular economy principles. Innovations in recycling, upcycling, and waste-to-energy technologies contribute to reducing landfill waste and promoting resource efficiency.

Biodegradable Packaging: Indigenous technologies can lead to the development of biodegradable packaging materials made from locally available resources. This addresses the issue of plastic pollution and reduces the environmental impact of packaging.

Community-Led Initiatives: Community-driven waste management initiatives, rooted in indigenous knowledge, empower local communities to take responsibility for their waste. Innovations in decentralized waste processing can be integrated into these community-led efforts.

4. Conservation and Restoration:

Biodiversity Monitoring: Indigenous knowledge about local flora and fauna contributes to biodiversity monitoring efforts. Combining traditional ecological knowledge with modern tools facilitates effective conservation and restoration of ecosystems.

Habitat Restoration: Indigenous technologies for habitat restoration, including traditional agroforestry and afforestation practices, promote biodiversity and enhance ecosystem resilience. These methods contribute to mitigating the impacts of deforestation.

Sustainable Fisheries: Indigenous fishing practices that align with sustainable harvesting methods and community-based management contribute to the conservation of aquatic ecosystems. Combining traditional knowledge with modern fisheries management enhances the sustainability of fishing practices.

5. Climate Change Adaptation:

Traditional Ecological Knowledge: Indigenous communities often possess traditional ecological knowledge that aids in climate change adaptation. This knowledge can inform strategies for resilient agriculture, water management, and disaster preparedness.

Natural Resource Stewardship: Indigenous technologies emphasize the sustainable use of natural resources, fostering resilience in the face of climate change. Practices such as traditional agroforestry contribute to soil conservation and water retention.

6. Eco-friendly Technologies in Industry:

Green Manufacturing: Indigenous technologies can inspire green manufacturing practices that minimize environmental impact. Innovations in eco-friendly materials, energy-efficient processes, and waste reduction contribute to sustainable industrial development.

Low-Carbon Transportation: Indigenous technologies can influence sustainable transportation solutions. From traditional modes of transport to innovations in electric and hybrid vehicles, eco-friendly transportation practices can reduce carbon emissions.

In conclusion, the integration of indigenous technologies into sustainable development practices offers a transformative pathway towards a future where humanity achieves balance and harmony with nature. Envisioning a future where these technologies play a pivotal role in addressing environmental challenges, the chapter underscores the importance of respecting traditional wisdom while embracing innovation for a sustainable and resilient world.

7.5 Nurturing the Next Generation of Innovators

The legacy of Indian science giants beckons the need for nurturing the next generation of scientists and innovators. This section explores strategies for fostering a culture of curiosity, creativity, and critical thinking among the youth. The chapter underlines the importance of educational reforms, mentorship programs, and collaborative initiatives to empower the future custodians of indigenous technologies.

Nurturing the next generation of innovators is crucial for ensuring a sustainable and progressive future. This involves cultivating a mindset of curiosity, creativity, and problem-solving among young minds. The chapter explores various aspects of fostering innovation in the next generation, including education, mentorship, and creating conducive environments for innovation.

1. Education for Innovation:

STEM Education: A strong foundation in Science, Technology, Engineering, and Mathematics (STEM) subjects is essential. Integrating hands-on projects, practical applications, and real-world problem-solving into the curriculum fosters innovation from an early age.

STEAM Approach: Including Arts in the STEM framework (STEAM - Science, Technology, Engineering, Arts, and Mathematics) encourages a holistic approach to innovation. It recognizes the importance of creativity and design thinking alongside technical skills.

Project-Based Learning: Project-based learning methodologies provide students with opportunities to apply theoretical knowledge to real-world projects. This approach encourages collaboration, critical thinking, and innovation.

2. Mentorship and Role Models:

Industry Collaboration: Establishing partnerships between educational institutions and industries allows students to interact with professionals, gain real-world insights, and understand industry needs. This collaboration can inspire innovation and guide career paths.

Mentorship Programs: Mentorship from experienced professionals and role models is invaluable for aspiring innovators. Mentorship programs connect seasoned individuals with young innovators, providing guidance, advice, and a supportive network.

Entrepreneurial Exposure: Exposing students to entrepreneurship experiences, such as startup challenges, pitch competitions, and internships, nurtures an entrepreneurial mindset. It encourages risk-taking and the pursuit of innovative solutions to real-world problems.

3. Creating Conducive Environments:

Innovation Hubs and Incubators: Establishing innovation hubs and incubators within educational institutions or as standalone entities provides a physical space for collaboration, experimentation, and the development of innovative projects.

Flexible Learning Spaces: Designing flexible and collaborative learning spaces fosters an environment conducive to innovation. Spaces that encourage teamwork, creativity, and hands-on exploration contribute to a culture of innovation.

Access to Technology: Ensuring students have access to cutting-edge technologies, tools, and resources is crucial. This includes laboratories, 3D printers, coding equipment, and software platforms that enable hands-on learning and experimentation.

4. Encouraging Diversity and Inclusion:

Diverse Perspectives: Encouraging diversity in educational settings ensures a variety of perspectives and approaches to problem-solving. Exposure to diverse viewpoints fosters creativity and innovation.

Inclusive Programs: Designing inclusive programs that cater to students with varied backgrounds and abilities ensures that innovation is not limited to a specific demographic. Inclusivity promotes a rich and diverse innovation ecosystem.

5. Global Learning and Collaboration:

International Exchanges: Facilitating international exchanges, collaborations, and partnerships exposes students to global perspectives. Interacting with peers from different cultures enhances cross-cultural understanding and stimulates innovative thinking.

Virtual Collaboration: Leveraging technology for virtual collaboration allows students to connect with peers, experts, and mentors globally. Virtual platforms enable the exchange of ideas, fostering a global network of innovators.

6. Soft Skills Development:

Communication and Presentation Skills: Developing effective communication and presentation skills is crucial for innovators. The ability to articulate ideas clearly and persuasively is essential for gaining support and funding for innovative projects.

Critical Thinking and Problem-Solving: Fostering critical thinking skills enables students to analyze problems, identify opportunities, and develop innovative solutions. Problem-solving skills are fundamental for navigating challenges in the innovation process.

7. Recognizing and Celebrating Innovation:

Innovation Awards and Competitions: Establishing innovation awards and competitions provides a platform for students to showcase their ideas. Recognition and celebration of innovative achievements inspire others and create a culture that values innovation.

Showcasing Success Stories: Sharing success stories of young innovators who have made a significant impact in their fields serves as inspiration for the next generation. Highlighting diverse pathways to success encourages varied approaches to innovation.

In conclusion, nurturing the next generation of innovators requires a multifaceted approach that encompasses education, mentorship, conducive environments, and a focus on soft skills. By fostering a culture that values innovation and provides the necessary support structures, society can empower young minds to become the innovators and problem-solvers of tomorrow.

7.6 Global Collaborations: Positioning India on the World Stage

The chapter concludes by highlighting the potential for global collaborations in the realm of indigenous technologies. By forging partnerships with international counterparts, India can leverage diverse expertise, share knowledge, and contribute to global advancements. The narrative emphasizes the role of India as a key player in the global technological landscape, offering insights and innovations that resonate on a worldwide scale.

Global collaborations play a pivotal role in positioning India on the world stage as a hub for innovation, research, and economic growth. This chapter explores the importance of international partnerships, collaboration frameworks, and the strategic advantages for India in engaging with the global community.

1. Research and Innovation Partnerships:

Knowledge Exchange: Collaborating with global research institutions facilitates knowledge exchange, allowing Indian researchers and innovators to access cutting-edge advancements and contribute to the global pool of knowledge.

Joint Research Initiatives: Engaging in joint research initiatives with international partners accelerates innovation. Collaborative projects address global challenges and provide opportunities for Indian researchers to work on interdisciplinary projects.

Technology Transfer: International collaborations enable the transfer of technology and expertise. Leveraging innovations from global partners enhances India's technological capabilities and supports the growth of industries.

2. Economic Alliances and Trade:

Market Access: Collaborations with foreign industries provide Indian businesses with access to global markets. Exporting indigenous technologies and products contributes to economic growth and strengthens trade relations.

Foreign Direct Investment (FDI): Building collaborative frameworks attracts foreign investment. Joint ventures and partnerships with international companies bring capital, technology, and market opportunities to India.

Global Supply Chains: Integration into global supply chains enhances India's position in the manufacturing sector. Collaborative efforts ensure that Indian industries are competitive and aligned with international standards.

3. Academic and Educational Ties:

Student Exchange Programs: Collaboration between Indian and foreign educational institutions through student exchange programs fosters a diverse and globalized education environment. It exposes students to different perspectives and enhances cross-cultural understanding.

Joint Degree Programs: Establishing joint degree programs allows students to earn qualifications from both Indian and international institutions. This enriches the educational experience and enhances the global reputation of Indian educational institutions.

Faculty Collaboration: Encouraging collaboration between Indian and foreign faculty members promotes research partnerships, cross-cultural learning, and the exchange of pedagogical best practices.

4. Diplomacy and Soft Power:

Cultural Exchanges: Collaborative cultural programs and exchanges contribute to soft power diplomacy. Sharing art, literature, and cultural heritage enhances India's global influence and fosters mutual understanding.

International Events and Conferences: Hosting or participating in international events and conferences strengthens India's diplomatic ties. It provides a platform for showcasing innovations, engaging with global leaders, and shaping international discourse.

Humanitarian and Global Challenges: Collaborating on humanitarian initiatives and addressing global challenges, such as climate change and public health crises, positions India as a responsible and proactive global player.

5. Technological Diplomacy:

Space and Aerospace Collaborations: Engaging in international space collaborations and aerospace projects enhances India's technological prowess. Collaborative space missions and satellite launches strengthen diplomatic ties and showcase India's capabilities.

Cybersecurity Cooperation: Collaborating on cybersecurity initiatives fosters a secure digital environment. Information sharing and joint efforts against cyber threats position India as a reliable partner in the global digital landscape.

Emerging Technologies: Partnerships in emerging technologies, such as artificial intelligence, blockchain, and quantum computing, contribute to India's leadership in the development and application of these technologies.

6. Multilateral Forums and Organizations:

Participation in Global Organizations: Active participation in international organizations and forums enhances India's diplomatic standing. Contributions to discussions on global issues and collaboration within organizations reinforce India's role as a responsible global actor.

Trade Agreements: Engaging in regional and global trade agreements opens avenues for economic collaboration. Negotiating favorable trade terms enhances India's competitiveness in international markets.

Climate Agreements: Collaborating on climate initiatives and agreements demonstrates India's commitment to environmental sustainability. Shared efforts to address climate change contribute to India's global reputation.

7. Science and Technology Diplomacy:

Strategic Collaborations in R&D: Forming strategic collaborations in research and development strengthens India's scientific and technological capabilities. Collaborative projects address shared challenges and contribute to global advancements.

Participation in International Consortia: Joining international consortia and collaborative research networks enables India to contribute expertise and benefit from shared resources. This enhances the impact of Indian research on the global stage.

Exchange of Best Practices: Participating in international forums allows India to share best practices in science, technology, and innovation. It fosters a culture of continuous improvement and learning from global experiences.

In conclusion, global collaborations are instrumental in positioning India on the world stage as a leader in innovation, diplomacy, and economic growth. By fostering partnerships across various sectors, India can leverage the strengths of the global community to address challenges, drive technological advancements, and contribute to a more interconnected and prosperous world.

Epilogue: A Vision of Developed India

As we bid farewell to this exploration, the epilogue paints a vivid picture of a Developed India propelled by cutting-edge indigenous technologies. The vision encompasses a harmonious blend of tradition and innovation, where technological advancements are deeply rooted in the nation's cultural fabric. The legacy of Indian science giants lives on, inspiring a future where indigenous technologies continue to thrive, contributing to the prosperity and well-being of the nation and beyond.

The epilogue envisions a developed India that has realized its potential as a global powerhouse, not only in economic terms but also in innovation, sustainability, and societal well-being. It paints a picture of a nation that has overcome challenges, embraced opportunities, and charted a path toward comprehensive development. Here's a glimpse of the vision for a developed India:

1. Economic Prosperity:

Innovative Industries: India stands at the forefront of innovation-driven industries, with a robust ecosystem supporting startups, research and development, and cutting-edge technologies.

Global Economic Influence: A developed India is a key player in the global economy, with strong economic ties, foreign investments, and a diversified export portfolio. The nation's economic policies prioritize sustainability and inclusive growth.

Smart Infrastructure: Modern and efficient infrastructure, including smart cities, advanced transportation networks, and sustainable energy systems, supports economic activities and enhances the quality of life.

2. Technological Leadership:

Advanced Research and Development: India leads in research and development across various disciplines, including artificial intelligence, biotechnology, space exploration, and clean energy.

Innovation Hubs: The country is dotted with innovation hubs, fostering collaboration between academia, industry, and government. These hubs serve as incubators for groundbreaking ideas and technological advancements.

Global Technological Solutions: Indian technologies and innovations address global challenges, earning the nation recognition as a reliable source of solutions in areas such as healthcare, climate change, and sustainable development.

3. Sustainable Development:

Environmental Stewardship: India prioritizes environmental conservation and sustainability. The nation actively participates in global initiatives, leading the way in combating climate change, preserving biodiversity, and promoting eco-friendly practices.

Green Energy Transition: The energy landscape has undergone a significant transformation, with a substantial share of renewable energy sources powering the nation. India serves as a model for sustainable energy practices and a leader in combating climate change.

Circular Economy: A circular economy approach is ingrained in industrial practices, waste management, and resource utilization. India pioneers innovative solutions for recycling, upcycling, and minimizing environmental impact.

4. Inclusive Society:

Quality Education for All: Accessible and high-quality education is available to every citizen, fostering a knowledgeable and skilled workforce. Education is tailored to nurture critical thinking, creativity, and a sense of responsibility.

Universal Healthcare: A robust healthcare system ensures universal access to quality medical services. Innovations in healthcare technologies and preventive measures contribute to a healthier population.

Social Equality: The society embraces diversity, promoting gender equality, inclusivity, and social justice. Policies and initiatives actively address disparities and ensure that every citizen has equal opportunities.

5. Cultural Renaissance:

Preservation of Heritage: India's rich cultural heritage is preserved and celebrated. Digital technologies and immersive experiences ensure the accessibility and continuity of traditional arts, languages, and practices.

Cultural Diplomacy: India's cultural influence extends globally, fostering positive international relations. Cultural exchanges, artistic collaborations, and showcasing the nation's diverse heritage contribute to soft power diplomacy.

Innovation in the Arts: The arts and creative industries flourish, with innovations in literature, music, cinema, and visual arts reflecting contemporary narratives and contributing to global cultural discourse.

6. Global Collaboration and Leadership:

International Collaboration: India actively collaborates with nations worldwide, contributing to global initiatives, peacekeeping efforts, and collaborative projects. The nation is recognized as a responsible and influential global player.

Technological Diplomacy: India's technological prowess is a cornerstone of its diplomatic efforts. Collaborations in space exploration, cybersecurity, and emerging technologies enhance the nation's global standing.

Humanitarian Leadership: India takes a leading role in humanitarian efforts, contributing to disaster relief, public health initiatives, and addressing global challenges with a sense of responsibility and compassion.

In conclusion, the epilogue paints a vision of a developed India that has harnessed its potential across economic, technological, social, and cultural dimensions. This vision is characterized by innovation, sustainability, inclusivity, and global leadership, setting the stage for a prosperous and harmonious future for the nation and its people.

.....***.....